Simply Good Physics 2: Electricity, Magnetism, and Waves

Arnold W. Yanof

January, 2014

Copyright © 2014 by Arnold W. Yanof

All rights reserved.

No part of this book may be reproduced by any means, nor transmitted, nor translated into a machine language or graphical or digital form without the written permission of the author.

To my Family ...
Past, present, and future generations

Contents

1 Electrostatic Force — 1
  1.1 Size of the Electrostatic Force — 1
  1.2 Comparison with Gravitational Force — 2
  1.3 Force of more than two charges — 3
  1.4 Energy Transfer due to Electrostatic Force — 5
    1.4.1 Potential Difference — 7

2 Insulators, Conductors and Transfer of Charge — 11
  2.1 Creating Static Charge — 12
    2.1.1 Creating Negative Static Charge — 12
    2.1.2 Creating Positive Static Charge — 12
    2.1.3 Everyday Ways to Create Static Charge — 12
  2.2 Movement of Static Charge Within a Body — 14
    2.2.1 Charges on Insulators — 14
    2.2.2 Charges on Metals: Spheres — 15
    2.2.3 Metal points — 16
  2.3 Transfer of Static Charge Between Bodies — 17
    2.3.1 Transfer by Contact — 17
    2.3.2 Transfer by Induction — 18
    2.3.3 Transfer by Breakdown — 19

3 Electrical Current, Resistance, and Power — 23
  3.1 Current and Circuits — 23
    3.1.1 Mechanical Analogy — 25
  3.2 Energy Flow in Circuits — 27
  3.3 Electrical Resistance: Definition — 28

3.4		Differences in Resistance	30
3.5		Ohm's Law	34
3.6		Advanced Circuits	36
	3.6.1	Circuits and Circulating Fluids	36
	3.6.2	Series circuits	37
	3.6.3	Parallel circuits	40
	3.6.4	More complex series-parallel circuits	44

4 The Electric Field   47

4.1		Defining The Electric Field	47
	4.1.1	Uniform Electric Field	49
4.2		Lines of Force	51
4.3		Rules for E-lines of Force	53
4.4		Parallel plate Capacitor	55
4.5		Gauss's Law	56
	4.5.1	Gauss's Law and Charged Spheres	57
	4.5.2	Parallel Plate Capacitor E-field	59
	4.5.3	Charge-voltage relationship for Capacitors	60
4.6		Practical Capacitors	61
	4.6.1	Energy Storage in a Capacitive Circuit	62
	4.6.2	Capacitor Dielectrics	63
	4.6.3	Timing In a Resistor-Capacitor Circuit	65
4.7		Capacitors in Series and Parallel	66

5 The Magnetic Field   69

5.1		Force between parallel Currents	72
	5.1.1	B-field Due to a Wire: Direction	73
	5.1.2	Righthand Rules	75
5.2		Force on a Charge Moving in a B-field	77
	5.2.1	Moving Charge Similar to Current	77
5.3		Current versus Moving Charge in B-field	78
	5.3.1	Angle Between B-field and Current Affects Force	80
5.4		Magnetic Torque on a Current Loop	82
5.5		Ampere's Law	83
	5.5.1	'Proving' Ampere's Law for a Straight Wire	85

		5.5.2 Solenoid B-field	85
	5.6	Magnetic Materials	87

6 Faraday's Law — 89
- 6.1 EMF due to Changing B-Flux 90
- 6.2 Limit On the EMF – the Self-Inductance 94
- 6.3 Inductance 95
- 6.4 Forces on Induced Currents – Magnetic Damping 95

7 AC Circuits — 99
- 7.1 AC Current 99
- 7.2 Current-Voltage Relationship in a Resistor 100
  - 7.2.1 Math Representation of AC 101
  - 7.2.2 AC Power in a Resistor 102
- 7.3 Current-Voltage in a Capacitor 105
- 7.4 Current-Voltage in an Inductor 107
- 7.5 Power in a Capacitor or Inductor 109
- 7.6 Transformers 110
- 7.7 Step-up and Step-down Transformers 114
- 7.8 Power in the Secondary 116
  - 7.8.1 Transformer Delivering Low Power 116
  - 7.8.2 Transformer Delivering High Power 117

8 Waves — 121
- 8.1 Waves on a Rope 122
  - 8.1.1 Frequency, Wavelength, and Wave Speed . 124
  - 8.1.2 Interference 125
  - 8.1.3 Wave Reflections 128
  - 8.1.4 Standing Waves and Resonance 129
  - 8.1.5 Modes of Wave Transmission 132
- 8.2 Sound Resonances 133
- 8.3 Electromagnetic Waves 135
  - 8.3.1 The Electromagnetic Spectrum 137
  - 8.3.2 Energy Content of an Electromagnetic Wave 138
  - 8.3.3 B-field Energy 141

9 Light and Optics — 145
- 9.1 Scattering and Reflection from Surfaces — 145
- 9.2 Metal vs. Dielectric Interfaces — 147
  - 9.2.1 Refraction and Snell's Law — 149
  - 9.2.2 Total Internal Reflection — 151
- 9.3 Polarization Effects of Reflection — 152
- 9.4 Convex Lenses and Images — 154
  - 9.4.1 Finite Source Size and Distance — 157
  - 9.4.2 Image Size Formula — 159
  - 9.4.3 Image Distance Formula — 160
- 9.5 Magnifying Lens Optics — 161
  - 9.5.1 Virtual Image of a Magnifying Lens — 161
  - 9.5.2 Magnifying Lens Image Size and Location — 162
  - 9.5.3 Combining Two Lenses: The Microscope — 164
- 9.6 Diffraction — 167
  - 9.6.1 Two Openings — 169
  - 9.6.2 Multiple Openings — 170
  - 9.6.3 Everyday Diffraction — 174

PREFACE

The familiar world of atoms, molecules, chemicals, liquids, gases, and solids is largely the result of the electrical interactions of positively and negatively charged particles. The taming of these charged particles in electric circuits has shaped the modern world during the 20th century. Electric circuits provide us with an extraordinary range of useful devices, including electric motors and electronic gadgetry. A major focus of Part 2 of basic physics is to develop an understanding of what makes these circuits and gadgets work.

Laboratory work accompanying this course can begin immediately to explore circuits, circuit elements, and power produced by electrical currents. Electrical currents consist of charges in motion. A basic understanding of moving charges, however, requires a parallel study of the forces between charges at rest, or 'static' charges. The forces between static charges determine how energy flows when charges move. Therefore the equations that begin this book focus on static charges; the resulting concepts are applied as soon as possible to 'currents' of moving charges.

The second important outcome of Part 2 will be an understanding of the creation, conversion, and storage of electrical energy. Generators are able to convert heat, water power, wind, and solar radiation to electrical energy. They operate with high efficiency. Batteries can store and release energy with little loss. Motors are able to convert electrical to mechanical power with high efficiency. The modern history of the world will be determined in large part by our ability to create, store, transmit and conserve electrical energy. Students need to understand the physics and be able to participate in the politics and the economics of electrical energy.

Chapter 1

Electrostatic Force

Objective: This chapter introduces the concepts and basic calculations for the force between electric charges. The electric force transfers electrical energy, just as mechanical forces produce kinetic energy and can store potential energy. The electric force also drives electric currents through conducting circuits. Circuits are introduced here and treated more thoroughly in a later chapter.

If there are two electrical charges Q_1 and Q_2 separated by a distance R, there is a force between them,

$$F = k\frac{Q_1 Q_2}{R^2} \tag{1.1}$$

Force F is measured in Newtons, N, Q_1 and Q_2 are measured in units C (Coulombs), and R is measured in meters. The constant k in this equation is very large, $k = 9 \times 10^9$, and its units are clearly N-m^2/C^2. If the signs of Q_1 and Q_2 are the same (both $-$ or both $+$), then F is positive, meaning the force pushes the two charges apart. If the signs of Q_1 and Q_2 are opposite, then F is negative, meaning they are attracted toward each other.

1.1 Size of the Electrostatic Force

The size of k is big because the Coulomb unit is a very large amount of charge, if it were free charge. Consider Q_1 and Q_2

each 1 C, and suppose the two charges are separated by 1 meter. Then the force between the two charges is $F_e = k*1^2/1^2 = 9 \times 10^9$ N. Compare this force with the weight of a 10 ton truck, which is about $F_g = Mg = 10^4$ kg $\times$ 9.8m/s^2 = 10^5 N. The force between Q_1 and Q_2 is $10^4 \times$ bigger than the force of gravity on a fully-loaded truck!

Why is the Coulomb so big? Because the charge flowing in electric circuits is also measured in Coulombs, and, as we shall see, large charges can flow in metal wires without building up much static charge and relatively small forces are sufficient to propel charges through wires.

In many household and laboratory electrostatics situations, the amount of charge is much smaller, say 100 nC (1 nC = 10^{-9} C). Then the force between two 100 nC charges separated by 1 cm would be $F_e = k * 10^{-7}$C $* 10^{-7}$C$/.01^2 = 0.9$ N. This is approximately the force of gravity on a 0.1 gram mass. The typical laboratory electrostatic charges and their corresponding Coulomb forces are quite small.

Although large static charges do not occur in ordinary lab environments, lightning from very large static charges does arise in Nature. The static charge build-up between storm clouds, and between the Earth and clouds, is approximately 10 C.

1.2 Comparison with Gravitational Force

Equation (1.1) is mathematically similar to the Universal Law of Gravitation, $F = GMm/R^2$. First of all, the force is an 'inverse square' law, meaning that it depends on $1/R^2$. For example, if the distance between charges doubles, the force between them decreases by a factor of 4, just like the force of gravity between two masses. Secondly, gravity depends on the product of the two masses, just as electrostatic force depends on the product of two charges.

The size of $G = 6.67 \times 10^{-11}$, however, is 20 orders of magnitude smaller than the electrostatic constant k. Therefore gravity is a much weaker force than the force between charges. A meaningful comparison can be made in the hydrogen atom. Hydrogen consists of a negatively charged electron circulating around a positively charged proton. The distance between proton and electron masses is of course identical to the distance between the proton and electron charges. The gravitational force in the hydrogen atom is

$$\begin{aligned} F_{H2,g} &= GM_p m_e/r^2 \\ &= 6.67 \times 10^{-11} * 1.67 \times 10^{-27} \text{kg} \\ &\quad * 9.11 \times 10^{-31} \text{kg}/(5.3 \times 10^{-11} \text{m})^2 \\ &= 3.6 \times 10^{-47} \text{N} \end{aligned}$$

By comparison, the electrostatic force, which actually holds the hydrogen atom together, is

$$\begin{aligned} F_{H2,el} &= k(e/r)^2 \\ &= 9 \times 10^9 * (1.60 \times 10^{-19} \text{C})/5.3 \times 10^{-11} \text{m})^2 \\ &= 8.2 \times 10^{-8} \text{N} \end{aligned}$$

Here $e = 1.60 \times 10^{-19}$ C is the charge on an electron. The electric force holding the atom together is $10^{39} \times$ greater than the gravitational force!

1.3 Force of more than two charges

If the force equation (1.1) shows only the interaction between two charges, how does it deal with three or more charges? To answer this, first recall Newton's 2nd and 3rd Laws. Newton's 3rd Law requires that whenever object #1 exerts a force F on object #2, then at the same time object #2 must exert an equal and opposite force on object #1. The electrostatic force certainly obeys this law– i.e. Equation (1.1) gives the same size force on Q_1 as on Q_2. Furthermore, the forces are indeed opposite in direction for the two charges Q_1 and Q_2, whether it is a case of attraction or repulsion.

Now remember the 2nd Law. In particular, Newton's 2nd Law states that the mass of an object times its acceleration is equal to the *sum* of the forces on the object. In the electrostatic case of three charges, Q_1, Q_2, and Q_3, we can find the sum of forces on Q_2 by adding the force between Q_1 and Q_2 together with the force of Q_3 on Q_2. So the equation for the sum of forces on charge Q_2 is

$$\vec{F} = k\frac{Q_1 Q_2}{R_{12}^2} + k\frac{Q_3 Q_2}{R_{23}^2} \tag{1.2}$$

This equation requires adding the two vector forces on Q_2, one being the interaction with Q_1, the other the interaction with Q_3. However, these forces are vectors so the two terms must be added in *vector* fashion.

Example of Charges in a Line

- Question: Let $Q_1 = Q_2 = Q_3 = 100$ nC, and let the separation between adjacent charges be 1.0 cm, as shown in Figure 1.1. What is the electrostatic force on Q_2? What is the force on Q_1?

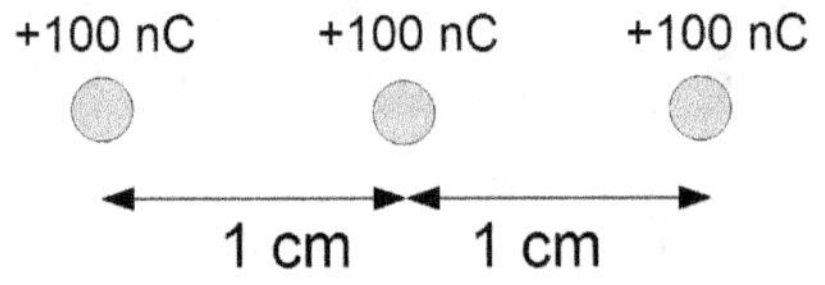

FIGURE 1.1
Three equal charges in a straight line

Answer: Even though each of the forces F_{12} and F_{32} is 0.9 N separately, in this geometrical configuration they push Q_2 in opposite directions. Together their vector sum cancels out and the net force on Q_2 is zero.

On the other hand, the forces on Q_1 due to Q_2 and Q_3 are additive. The force of Q_2 on Q_1 is again 0.9 N, now directed to the left; and the force of Q_3 on Q_1 is 4× smaller or 0.225 N, because $R_{13} = 2 \times R_{12}$. Hence the net force on Q_1 is 1.125 N, directed to the left.

Electrostatic Force

Example of Charges out of Line

- Question: What is the force on Q_2 if the charges are not in a straight line, but rather are arranged as in Figure 1.2. Assume again that $Q_1 = Q_2 = Q_3 = 100\,\text{nC}$, and let the separation between adjacent charges be 1.0 cm.

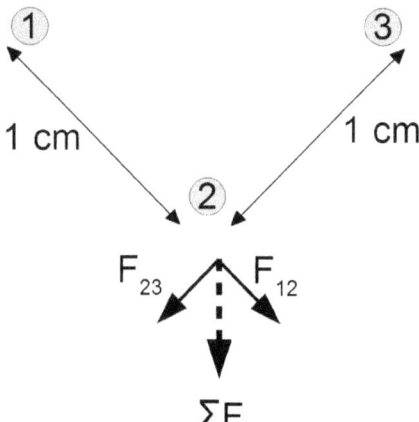

FIGURE 1.2 Three equal charges not in a straight line, giving southward resultant force on charge Q_2

Answer: There is now a non-zero net force on charge #2, due to vector addition of the forces from the other two charges. Charge Q_1 is to the northwest, and Q_2 is to the northeast, of Q_2. As above, the magnitude of each of the two forces on Q_2 from Equation (1.1) is $F_{12} = F_{23} = 0.9$ N. However the forces must add as vectors, so the two x-components cancel. The two y-components add, so the net force is

$$\sum F = 2 * F_{12} * \cos 45° = 1.27\,\text{N}$$

and the net-force direction is South.

1.4 Energy Transfer due to Electrostatic Force

We have seen in *Physics 1*, that whenever a mechanical force pushes or pulls an object, and the object moves in the direction of the force, then energy is transferred to that object. $W = \vec{F} \cdot \vec{\Delta x}$, where W is the energy, $\vec{F}$ is the force and $\vec{\Delta x}$ is the movement produced by the force. This is naturally true for charged particles as well. Take for example a typical alkaline battery. It has a small positive terminal at the top and a negative terminal at the bottom. (Actually, the negative terminal of the battery is usually a cylindrical can that extends up the sides and is insulated from the top, but for the sake of simplicity, consider the negative terminal to be just the bottom of the battery.) At the bottom there are a number of electrons contributed to the terminal by the chemical action inside the battery. At the positive terminal there is a deficit of electrons taken away by the chemicals from an electrode inside the battery attached to the top terminal. Therefore a free electron outside the battery at a height in between the bottom and the top has a force on it which will drive it toward the positive terminal. See Figure 1.3

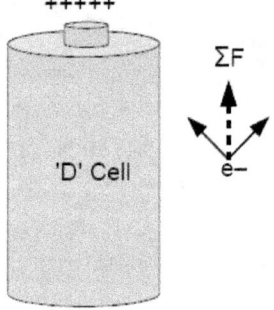

FIGURE 1.3
A free electron experiences a force that pushes from the − to the + battery terminal

Now what happens if the electron moves? The electron could:

1. gain potential energy by being lowered from the + to the − terminal (just as a mass m could gain potential energy (PE) by being lifted up against the force of gravity). Or

2. the electron could be let free to move in the direction of the force and gain kinetic energy. Still a third option is

3. the electron may move with the force, lose PE, and give up the energy as heat and light in an electric device like a lamp

(just as a box can slide down a ramp and give up its *mgh* potential energy to frictional heating).

Each of the above motions results in a change in energy. The change in energy is very important to us, and a special concept, the *voltage*, facilitates energy calculations when electric charges move. The voltage at any particular point in space gives the potential energy of a unit of positive charge if it were located at that point. As with gravitational potential energy, the differences in potential energy are what is important. These potential differences are discussed in the next section.

Important Principle of Potential

- A positive charge gains PE when it moves toward + charge, and loses PE when it moves toward − charge. A negative charge *loses* PE when it moves toward + charge, and *gains* PE when it moves toward − charge.

1.4.1 Potential Difference

The battery has the characteristic property, due to its chemical nature, that almost regardless what amount of electrons flow around the circuit, each electron gets the same energy as it goes from the bottom to the top of the battery. In fact, if 1 Coulomb of electrons passes from the bottom to the top of a 1.5 volt battery, the total energy going to the electrons is 1.0 C × 1.5V = 1.5 Joules. Similarly, if 2.0 C of electrons go from the bottom to the top, the total energy going to the electrons is 3.0 J, and so forth. The energy produced per Coulomb of charge is called the *potential difference* or the *voltage* of the battery. An alkaline battery has a voltage of 1.5 V (Volts).

In general, the PE stored when charge Q is pushed from a location with voltage V_1 to a new location with voltage V_2 is

$$\Delta U_{PE,el} = (V_2 - V_1) * Q = \Delta V * Q \text{ (Joule)} \quad (1.3)$$

A rearrangement of Equation 1.3 defines the voltage change in going from position 1 to position 2 as the potential energy stored for each Coulomb of charge moved from position 1 to position 2:

$$\Delta V = \Delta U_{PE,el}/Q \text{ (Volts)} \quad (1.4)$$

PE and Voltage Examples

1. Question: How much kinetic energy does a free electron get from the battery when it goes through an external path or circuit from the bottom $(-)$ terminal of the battery to the top $(+)$ terminal? You are given that the charge of the electron is 1.60×10^{-19} C and the battery voltage is 1.5 V.

    Answer: When a charge of -1.60×10^{-19} C goes from a lower voltage to a more positive voltage, through a potential difference of 1.5 V, it loses potential energy and gains kinetic energy. $\Delta U_{PE} = Q \Delta V = -1.60 \times 10^{-19}$ C $* 1.5$ V $= -2.4 \times 10^{-19}$ J. This is the PE loss.

    If the electron were not free, but part of a circuit, the load in the circuit absorbs this energy. For example, the circuit load could be the heating element of an appliance, and the electron's energy given up to create heat energy.

2. Question: A charge $Q = +6$ C passes from the positively charged upper thunder cloud to the negatively charged lower part. See Figure 1.4. (a) Why is the voltage higher at the top of the cloud? (b) If the energy released by the lightning bolt is 400 Mega joule, or 4.0×10^8 J, what is the voltage difference between the top and bottom of the cloud?

    Answer: (a) The voltage is higher at a concentration of positive charge, because it takes energy to bring each additional $+$ charge to that location. (b) The potential energy stored in the cloud is converted into heat, light, sound, and mechanical energy by the lightning bolt. Equation (1.4) says the voltage difference is equal to the stored potential energy

Electrostatic Force

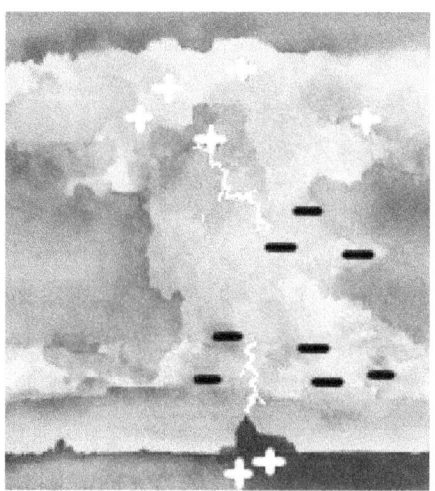

FIGURE 1.4
The friction of air circulation causes build-up of large static electrical charges in thunderclouds

divided by the amount of charge:
$\Delta V = U/Q = 4.0 \times 10^8 \, \text{J} / 6.0 \, \text{C} = 67 \times 10^6 \, \text{V}$
(This assumes that discharging the 6 C does not substantially reduce the voltage difference during the lightning strike.)

3. **Question:** In an x-ray machine, electrons accelerate in a vacuum tube from a heated wire at voltage $V_1 = 1000 \, \text{V}$ to an 'anode' or positive plate at voltage $V_2 = 4000 \, \text{V}$. (a) What is the kinetic energy KE of the electron when it strikes the anode, and (b) what is its velocity, v? The mass of the electron is $m_e = 9.1 \times 10^{-31} \, \text{kg}$

   Answer: When a negative charge of $1.60 \times 10^{-19} \, \text{C}$ moves to a more positive voltage, it gains energy. In a vacuum, the electron can accelerate freely, and all the energy goes into KE. The amount of energy is, according to 1.3,
   (a) $KE = U = Q\Delta V = 1.60 \times 10^{-19} \, \text{C} * 3000 \, \text{V} = 4.8 \times 10^{-16} \, \text{J}$ and
   (b) $v = \sqrt{2KE/m_e} = 3.25 \times 10^7 \, \text{m/s}$.

Chapter 2

Insulators, Conductors and Transfer of Charge

Objective: This chapter explores the phenomena of excess charges on various objects. The purpose is to see how excess charge is created, and analyze the forces between charges to predict how charge transfers from one object to another, or flows through an object.

In Chapter 1, Equation (1.1) described the force between two charges. There are many materials, such as wood, dry air, cloth fibers, plastics, and others which prevent a charge from moving in response to the force it feels. These materials are called *insulators*. On the other hand, metals, ionized gases, and liquid ionic solutions are examples of materials which allow charges to move freely under the influence of electric forces. These materials are called *conductors*.

The passage of charges from one place to another is known as electric *current*. On the other hand, when insulators prevent the flow of charge, positive and negative charge can build up and not move, move very slowly, or move sporadically. This charge is known as *static* charge.

2.1 Creating Static Charge

When insulating materials rub together, electrons get scraped off one material and stick to the other material. The most common experience of this occurs when one shuffles across a carpet or slides across a car seat on a dry day. In humid climates, this may not be a typical experience. However, a clothes dryer is always dry enough that the friction between different fabrics or between the fabrics and the parts of the dryer cause the formation of static charge, once the clothes are dry. The attractive force between different types of cloth is very apparent, and is just Equation (1.1) at work. Positive charge is left on some types of cloth, excess electrons accumulate on other types, and the two articles of clothing then attract each other.

2.1.1 Creating Negative Static Charge

The traditional way to make a negatively charged object is to rub a hard rubber rod with cat fur. See Figure 2.1. The rod becomes negatively charged, and the fur gets a positive charge. A wide variety of material pairs exhibit this behavior. Many plastic or softer rubber materials can substitute for the rubber rod, even your own skin. Human hair works as well as cat fur.

2.1.2 Creating Positive Static Charge

The traditional way to make a positively charged object is to rub a glass rod with silk. The glass rod becomes positively charged, and the silk negative. See Figure 2.2.

2.1.3 Everyday Ways to Create Static Charge

Many common types of friction cause static charge to develop. One of the most dangerous is the flow of flammable materials out of a container through a pipe or tube. When gasoline or other non-conducting fuel is pumped through a tube, a static charge

Insulators, Conductors and Transfer of Charge

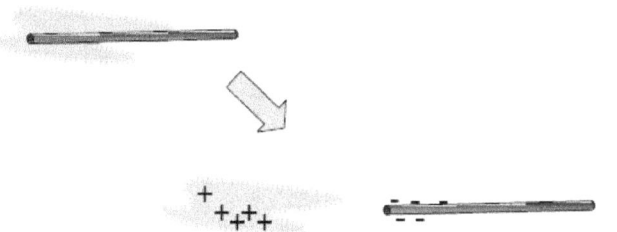

FIGURE 2.1
A rubber rod rubbed with cat fur becomes charged with electrons. The drawing above left indicates the rubber rod being rubbed by the fur. Then below right the fur and the rod are separated, thereby separating + and − charges.

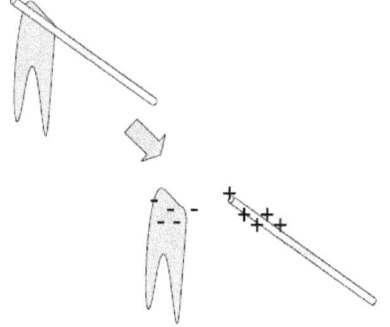

FIGURE 2.2
A glass rod rubbed with silk becomes charged positively

builds. This can be extremely destructive if a spark discharges in the presence of the flammable liquid. For this reason, trucks frequently have a discharging mechanism. That is, the rubber tires permit a static charge to build up on the vehicle, even while fueling, so a conducting belt makes contact with the ground to discharge the vehicle safely. You may have experienced the build-up of static charge due to the flow of air. If you operate a leaf blower, paint sprayer, or vacuum, for example, you can experience a shock due to the separation of charge between the nozzle and the flowing liquid, gas or fine solid particles rushing through it. Another dangerous situation is the flow of grain out of a chute. The static spark can readily ignite an explosion in the powdered grain, which is highly flammable because it is finely divided.

2.2 Movement of Static Charge Within a Body

Once an object has a static charge, or if different parts of an object hold different charges, the charge would like to move under the influence of mutual electrical forces. If the charges can move under the influence of the mutual force between different charged regions, this will lower the potential energy. This is completely analogous to an elevated mass lowering its potential energy by falling downward toward the Earth under the force of gravity. For example, two negatively charged regions can lower their mutual potential energy by moving away from each other.

2.2.1 Charges on Insulators

Imagine if various $+$ and $-$ charges are placed in different locations on or within an insulating material - i.e., one which does not readily conduct electricity. Although there are forces due to Equation (1.1) between them, they remain immobile. If the material is a good insulator, and depending on temperature and humidity conditions, the charges can remain fixed in position for a considerable length of time. This is illustrated in Figure 2.3.

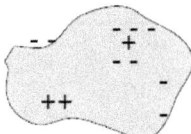

FIGURE 2.3
Charges placed on or inside an insulating body are fixed

2.2.2 Charges on Metals: Spheres

On the other hand, if various + and − charges are placed on different locations on or within a piece of metallic material, the charges move quickly because of the forces of Equation (1.1) between them. The like charges will repel, but + and − charges will attract one another and move toward each other and neutralize. In the end, some net positive or negative charge remains on the sphere, depending on whether there was more + or more − charge placed on the object to begin with.

Assume the net charge on a metal sphere is negative. In metal, the electrons are free to move from place to place. Therefore the electrons repel one another and tend to spread across the metal. Furthermore, the − charge does not remain inside the metal sphere, but rather tends to spread over its surface. The reason is any two negatively charged regions in the metal will move apart as far as possible. The best separation between the negative charges is achieved if they spread uniformly over the surface of the sphere.

On the other hand, assume the net charge on the metal sphere is positive. Obviously the positive charge will attract some of the electrons in the metal sphere toward it. However, as electrons move toward the positive charge, they leave behind positive metal ions which cannot move freely. Therefore in effect the + charge also spreads apart across the surface of the sphere, until the charge is again spread uniformly over the surface. This is illustrated in Figure 2.4.

FIGURE 2.4
Charges on a metal sphere move until they are uniformly distributed over the area of the surface

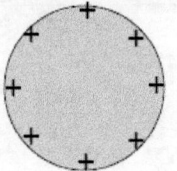

2.2.3 Metal points

If there is a point or protuberance on a metal object, mutual repulsion will cause a disproportionate amount of charge to push toward the pointy end. Figure 2.5 shows a charged metal sphere that has been elongated in shape towards one side. The charge on

FIGURE 2.5
Charges on a non-spherical charged metal object distribute themselves unevenly, accumulating higher charge density near a point.

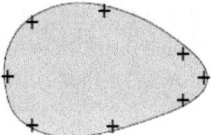

the elongation will be repelled by the like charge on the rest of the sphere, and the protuberance allows some charge to move away from the main body under the influence of this force. The result is a higher surface charge density (C/m^2) on the pointed end. This is very significant, because as the protuberance becomes pointier, the charge density gets proportionally more concentrated at the point. At a very sharp point, there is a small total amount of charge; but it is piled up into a very small, very concentrated region. So any ions coming near the point can experience an extremely large force due to (1) the *closeness*, i.e., due to the $\approx 1/R^2$ factor in Equation (1.1); and (2) the higher charge density at

Insulators, Conductors and Transfer of Charge

the point. We will see in the following section that the very large electrostatic forces near a charged point can cause breakdown— e.g., arcing— through the surrounding air, which is normally an insulator.

2.3 Transfer of Static Charge Between Bodies

Movement of static charge within a body was discussed in the previous section. Here the discussion extends to movement of charge from one body to another.

2.3.1 Transfer by Contact

If you touch a negatively charged rubber rod to a metal bar, this gives electrons on the rubber rod an opportunity to get farther apart by transferring to the metal. Similarly, when a charged metal object touches another metal object, electrons move so that the excess charge will spread over both.

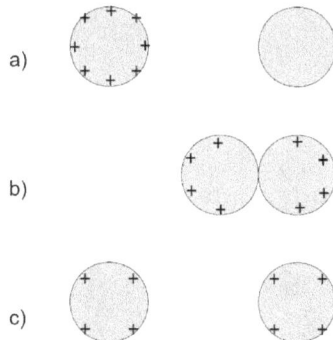

FIGURE 2.6
How charge redistributes in a transfer by contact between two metal spheres.

Transfer by Contact Example

- Question: A metal sphere is negatively charged. It approaches (a), contacts (b), and withdraws from (c) a second sphere which is initially uncharged. Show the distribution of charge at each step.

  Answer: The charge distributions are as shown in Figure 2.6.

2.3.2 Transfer by Induction

This is a method of using a charged object to transfer the opposite-sign charge to a neutral object. If you bring a negatively charged rubber rod near a metal bar, the − rod repels electrons in the bar. The electrons in the metal bar will go to the far end of the bar, and leave behind some + ions, which are not free to move in the metal. It may appear as though the negatively charged rubber rod *attracted* the + charges in the metal, and there *is* an attractive force between the rubber rod and the metal. However, the + charges actually cannot move. The situation will appear as in 2.7

FIGURE 2.7
If a charged rubber rod is placed in proximity to a metal bar, the rod pushes electrons toward the opposite end of the metal bar.

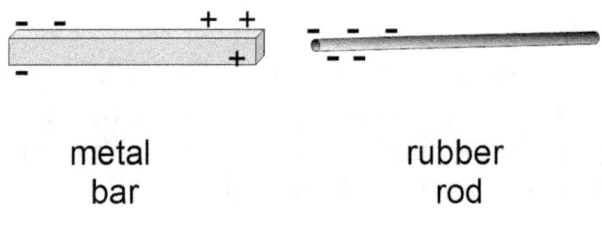

If you hold the metal object in your hand while bringing it near a negatively charged rod, the electrons in the metal will again be pushed away from the negatively charged rod. Now that your hand is on the metal, the negatively charged rubber rod can push some electrons all the way out of the metal object and into your body (and down to the floor, ... *etc*). Thus the electrons can get even further away from the rubber rod. Figure 2.8 shows this happening.

Explain Why

1. Question: Why is there an attractive force between the rubber rod and the metal bar in Figure 2.7?

   Answer: There are three regions of charge in the picture: the charge on the rubber rod $-q_{rr}$, the charge $+q_{bar}$ induced

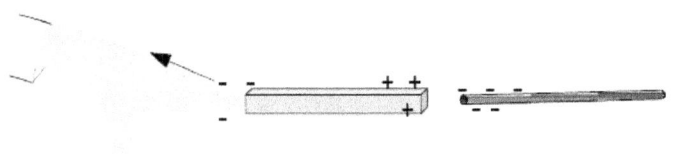

FIGURE 2.8
The hand provides a path for some electrons to leave the metal bar. Withdrawing the hand then leaves the metal with an *induced* + charge.

on the metal bar, and the charge $-q_{bar}$ repelled away to the far end of the metal bar. Considering the Coulomb's Law forces on the metal bar, the products $-q_{rr} \cdot q_{bar}$ and $+q_{rr} \cdot q_{bar}$ are equal and opposite because the bar started out, and remains, electrically neutral. However, the (+) charge is closer to the rubber rod, so the $1/R^2$ law makes attraction to the (+) end of the metal rod greater than the repulsion between the rubber rod and the (−) end of the bar.

2. Question: Show what the charge will look like on the bar in Figure 2.8 if the hand is first withdrawn, and then the rubber rod is pulled away.

2.3.3 Transfer by Breakdown

A third method by which charge often transfers is electrical breakdown. For example, if you rub a balloon on your hair and then bring the balloon close to your finger, in a darkened room, you will see a spark jump between your finger and the balloon.

Consider the following sequence of events leading to this spark, also called a *discharge*:

1. The portion of the balloon that was rubbing your hair becomes strongly negatively charged.

2. You may hold the other end of the balloon without draining any of this charge because the balloon is a good insulator.

3. Your hair becomes positively charged, but gives up much of its charge to your body, which stays positively charged if your shoes have rubber soles and the air is fairly dry.

4. Your finger is somewhat pointy, so a disproportionate amount of + charge builds up on the tip of your finger as it approaches the balloon.

5. As your finger approaches the balloon, the force between the − balloon and the + finger becomes quite large. The balloon may bend toward your finger tip or, if floating in the air, may move toward your finger, thereby increasing the attractive forces even more.

6. Furthermore, the forces on an ion or other charged particle in the air passing through the gap between finger and balloon also become very large. This will cause the ion or small particle to accelerate.

7. The accelerated charged particle bumps into other molecules in the air, ionizing the latter.

8. The ions accelerate and create other ions by bumping into air molecules. The ions multiply into an avalanche of charge, creating sound, heat and light visible as a spark between the balloon and your finger.

On a much larger scale, a similar sequence of events can cause lightning to strike between a cloud and a pointy object on the ground, such as a house, tree, or golfer. If the house has a lightning rod, the pointy tip of the lightning rod can produce a flow of charge into the atmosphere which neutralizes the charge in the atmosphere, before it has a chance to build up to a very large amount of charge, so no lightning bolt occurs. Or even if lightning strikes the lightning rod, the rod can conduct the charge safely between the ground and the atmosphere, bypassing the house itself.

Chapter 3

Electrical Current, Resistance, and Power

Objective: We have previously described the movement or transfer of static charge. This chapter explores the steady flow of charge, called <u>current</u> in electrical circuits. You will find there is a definite relationship between current and the battery voltage, known as <u>Ohm's Law</u>. Just as movement of a single charge under electrical forces transfers energy, you will see how a steady current transfers energy at a continual flow-rate, known as <u>power</u>.

3.1 Current and Circuits

Current is defined as the amount of charge per second moving through an object, such as a metal wire, a portion of the atmosphere, or even a vacuum chamber. The units of current are Amperes (A). One ampere is the current flowing when one Coulomb passes a specific point in the object in one second:

$$I = \Delta Q / \Delta t \qquad (3.1)$$

Examples

1. Question: If 10^9 electrons enter a wire every second, how much current is this, expressed in amperes?

   Answer: From Equation 3.1, $I = 10^9 \cdot 1.6 \times 10^{-19}\,\text{C/s} = 1.6 \times 10^{-10}\,\text{A}$

2. Question: If a lightning bolt discharges 6 C in 0.014 seconds, what is the average current in the lightning bolt during that interval of time?

   Answer: From Equation 3.1, $I = 6\,\text{C}/.014\,\text{s} = 428\,\text{A}$

Chapter 2 described how static charge moves under the influence of the force between the charges. Metals conduct charges very rapidly under the influence of electrostatic forces. Such currents are momentarily very large when static charge is transfered to a metal object, but decrease quickly. Insulators conduct poorly, and charges can move only very slowly, so the current due to static charge in insulators is very small. In metals or insulators, the movements of static charge are not steady. Rather, after a certain amount of time, the charge reaches an equilibrium such that virtually no further movement of charge occurs, and the current falls to zero.

A complete path of metal wires and useful electrical devices is called a *circuit*. Chapter 1 discussed how a battery can supply charge to move through a circuit on a steady basis. As long as the chemical action inside the battery continues, there is a steady supply of electrons on the − terminal of the battery. These electrons flow in a constant current through wires, lamps, calculators, toys, or various other electronic devices connected in the circuit and then re-enter the battery at the + terminal. The discussion in this chapter will deal entirely with these steady currents, called DC or *direct currents*.

3.1.1 Mechanical Analogy

It is useful to compare physical quantities in a 'mechanical cycle' with similar quantities in a battery circuit. An illustration of a mechanical cycle and a similar battery circuit is shown in Figure 3.1.

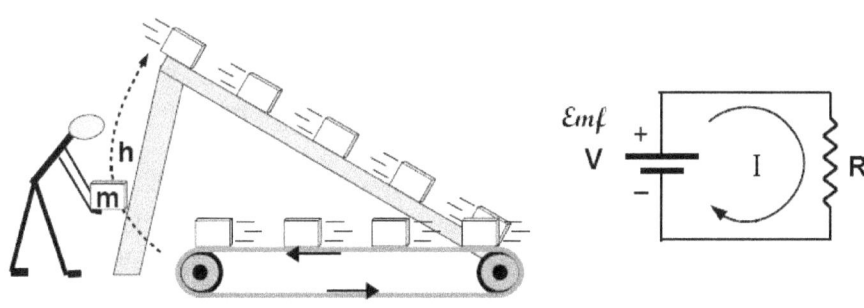

FIGURE 3.1 (left) In the mechanical cycle, a man raises boxes of mass m continuously onto a ramp where they slide down with friction. (right) In the analogous circuit, a battery supplies continuous current I to a load resistance R.

In the mechanical cycle in that figure, a number of boxes of mass m slide steadily down a ramp, giving up their potential energy to friction as they go down. At the bottom of the ramp they drop onto a belt which conveys them back to the man. We assume the belt does not add any net PE to the boxes, just brings them back to the man at the same height they were let off the ramp. We assume also the KE does not change throughout the cycle - i.e., speed of the boxes is constant.

Similarly, in the electrical circuit in the same figure, the battery is the source of energy. It is symbolized by short and long parallel lines. The short battery line is the negative terminal. The electrons flow out of the negative terminal. The straight, black lines symbolize wires. These convey the electrons from the battery to the resistive load, labeled R, and then back to the battery. The load R is the part of the circuit where electrical energy will

Description	Mechanical Cycle	Electrical Circuit
Energy 'container'	mass	charge
Container size	m, Kg	Q, Coulomb
Energy boost factor	$g * h$	V, Volt
Energy per container, J	mgh	QV
Power, Watt	mgh/T	$QV/T = \mathbf{I} \cdot \mathbf{V}$

TABLE 3.1: Comparison between the mechanical cycle and a similar electrical circuit

be destroyed and turned into heat, light, radio signals, etc. R represents a useful device such as a heater, lamp, cell phone, etc.

The electric current is symbolized by the clockwise arrow and labeled I. The convention is to think of currents in the direction of + charge flow. When an electron takes a step counter-clockwise in this circuit, it is equivalent to a + charge taking a clockwise step. This is because when the electron takes a step, it leaves a + charge where it came from and neutralizes a + charge that had been in front of it. It is just as though an equal + charge moved in the opposite direction. That is why the convention is to point the arrow in the direction opposite the flow of electrons, and it is a little easier for most engineers to think in terms of positive current flow.

Just as each box gains PE as it is raised by the man, then loses energy as heat as it slides down the ramp; in the same way, the electron gains PE as it goes downward through the battery, then loses its PE as it goes through R. Or, in terms of 'positive' current, the + charge increases PE as it goes up through the battery. Its energy remains constant as it flows through the wire, and then it loses its potential energy as it goes through R.

The mathematics of energy flow is similar for these two example systems, and is shown in Table 3.1. For the mechanical cycle, the thing that possesses the energy is the mass, measured in kilograms, whereas for the electrical circuit, the energy containing entity is the charge, measured in Coulombs (C). The unit of elec-

Electrical Current, Resistance, and Power   27

trical power is, as for mechanical power, the Watt = 1 Joule/Sec. In table 3.1, just note for now the formula for power in electrical circuits is $P = IV$ (Watt). We will discuss power in more detail in the following section.

Examples

1. Question: In the mechanical cycle shown in Figure 3.1, suppose the boxes weigh 1 N each, the height $h = 1.5$ m, and the man loads 1 box every 4 seconds. How much power dissipates as heat on the ramp?

   Answer: The man raises PE of each box $(mg)h = 1\,\text{N} \cdot 1.5\,\text{m} = 1.5\,\text{J}$. This is accomplished in T = 4 seconds, so the power produced by the man and dissipated on the incline is $P = (mg)h/T = 1.5\,\text{J}/4\text{s} = 0.38\,\text{W}$

2. Question: In the circuit shown in Figure 3.1, suppose the battery has an $EMF = 1.5$ Volt and drives a current of 0.25 amperes through the heater R. (In other words, 1 C of charge every 4 seconds.) How much power is dissipated in R?

   Answer: The power dissipation is given by Equation 3.2. It is $P = IV = 0.25\,\text{A} \cdot 1.5\,\text{V} = 0.38\,\text{W}$.

3.2 Energy Flow in Circuits

Many circuits are designed to deliver power from an energy source to an electrical load. As an introduction to electrical energy flow, consider the flow of energy in the mechanical analog, Figure 3.1. Suppose the man raises each box a height h. He thereby raises the PE of the box an amount mgh. If he picks up another box every T seconds, then the average power he produces is $P = mgh/T$ Watts. As the box slides down the ramp, it loses all its PE and converts it to heat. The average power dissipated in the ramp is therefore the same, mgh/T Watts. There is a continuous flow of energy created by the man and dissipated by friction.

Now for the circuit. The battery raises the PE of a charge Q by an amount QV, where V is the voltage of the battery. If the amount of charge Q circulates through the battery every T seconds, the average power produced by the battery is QV/T Watts. The electrical variables in this formula can regroup as follows:

$$P = \frac{Q}{T}V = IV \text{ (Watt)} \qquad (3.2)$$

This is a very useful equation. It gives the electrical power in a circuit as the current flowing times the voltage driving the circuit. This is the power produced by the battery, and is also the power that is dissipated in the electrical load R.

Power Examples

- Question: You are talking on a cell phone that has a 2.5 Volt battery. If the current for the cell phone circuits is 0.3 A, what is the power used by the cell phone?

  Answer: According to Equation (3.2), the power is 0.3*2.5 = 0.75 W. After talking for a while, the phone gets warm.

- Question: An electric car has a car battery that provides 218 Volts. While cruising, the current drawn from the battery is 32 Amps. How much power is this in Watts? In horsepower?

  Answer: The power is 32 A * 218 V = 6976 Watts. To convert, 6976 Watts × (1 hp/746 Watts) = 9.4 hp.

3.3 Electrical Resistance: Definition

It is always possible to add one battery on top of another to increase the voltage. If the battery in an electrical circuit is changed to a higher voltage, this increases the potential energy increase of an electron going through it. This typically causes larger current to flow through the load. Consequently, more charge at higher energy goes into the load, and it dissipates more power according to Equation 3.2. A familiar example is a 'brown-out', in which electrical lighting dims due to a temporary decrease in the voltage and current in the lamp. Another example is battery operated tools and toys which fall off in speed due to decreasing battery voltage near the end-of-life or low charge in the battery. In both these cases, a drop in the source voltage causes a decrease in current. Smaller current and smaller voltage result in a dramatic drop in power. When the voltage increases again, or the battery is recharged, the power comes back up to normal.

The fact that electrical current through a load generally increases with increasing voltage leads to our characterizing the load by the ratio between the voltage and the current. This ratio between voltage and current in a load is called the load *resistance*, R. By definition,

$$R = \frac{V}{I} \tag{3.3}$$

The unit of resistance is the ohm, symbolized Ω, and one ohm corresponds to a resistance such that 1 volt causes 1 ampere to flow.

Resistance Examples

- Question: A typical hair dryer uses 10 A when plugged into a 120 V wall outlet. What is the resistance of the heating coil in the hair dryer?

  Answer: The resistance is 120 V / 10 A = 12 Ω.

- Question: The resistance of dry skin is about 50,000 Ω. If you put your fingers across the terminals of a 9 V battery, how much current flows?

  Answer: Assume the electricity must pass through two layers of skin to enter and leave your finger. Then the combined resistance is $2 \times 50,000 = 100{,}000$ Ω. I = V/R = $9/10^5 = 9 \times 10^{-5}$ A.

3.4 Differences in Resistance

As discussed in Section 3.3, the resistance $R = V/I$ can characterize an electrical load, where V is the voltage applied, I is the current through the load, and R has the units of ohms. Different loads can have very different resistances. The factors influencing resistance are as follows:

- Pure metal loads have very low resistance. The inner atomic shells are tightly bound to the atom, so are not mobile enough to participate in conduction. You may recall from chemistry class that each metal atom has one or two valence electrons which readily separate from the atom, leaving an ion. A typical ionization reaction for a metal atom M is

$$M \rightarrow M^+ + e^-$$

In a metallic object, all the atoms in the object share all the valence electrons in common. A very slight voltage difference between locations in a metal object causes the electron pool to slide freely so as to neutralize the voltage difference. That is, negative charge will reduce its PE by moving toward positive voltage location, and any positive charge will attract electrons so as to reduce their PE.

The situation is analogous to your swimming pool. There is plenty of water in the pool - like the large number of charge carriers having high mobility in the metal. Suppose you briefly raise the water above the general level at one spot by pumping some water in at that spot. The water 'seeks its own level' by flowing rapidly away from the higher level. A flat surface is quickly restored. This analogy is illustrated in Figure 3.2. Hence in a metal, a small voltage difference can produce a large current, and this means the resistance expressed in ohms is a very small number. For example, the typical length copper wire for household items like a lamp cord has a resistance of $25 \times 10^{-3} \, \Omega$.

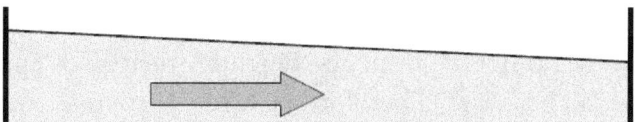

FIGURE 3.2
A metal is like your swimming pool; any height difference results in a big flow of current to restore surface level.

- Many materials have much higher resistance than pure metals:

  1. Insulators

     Insulators have almost no electrons that can move freely. In many insulators - e.g., many plastics, wood, glass, and others - the chemical bonds between neighboring atoms use up all or most of the valence electrons to form covalent bonds. This type of bonding localizes the electrons between pairs of atoms. Thus the atoms of an insulator, unlike a metal, do not share a common pool of valence electrons throughout the object.

     The water analogy for an insulator is a completely filled, closed, horizontal pipe. Raising one end will increase the PE of the water at that end. No horizontal flow occurs, however, because the higher potential water has no way to decrease the PE by flowing to the lower end.

  2. Impure metals and alloys

     Alloys are metal mixtures that have plenty of charge carriers, but their mobility is smaller than in a pure metal. Mixing metals often produces a material that consists of many small grains. Within a grain of pure metal, the crystal structure is very orderly. Electrons can go for a long distance without being stopped by a collision. In most alloys or impure metal, the electron hits a grain boundary or a crystal defect after a short distance. It then bounces off in a new, random direction different from the overall direction of current flow in the wire. Therefore electrons progress more slowly in response to a voltage difference. Many heating elements used in stoves, hair dryers, electric blankets, etc., use alloy wire to increase the resistance of the electrical load.

Modifying our swimming pool analogy to describe alloy conduction, consider a pool that is filled with obstacles such as loosely-packed stones together with the water. Now the water will still seek its own level in response to a height change or tilt at one location in the pool. However, the water flow experiences more drag, because it must flow through small and circuitous pathways between the stones.

3. Semiconductors

    Semiconductors have highly mobile charge carriers provided the temperature is high enough. However, they have much fewer carriers than metals. We will see in a later chapter that the carriers can be either electrons or the positive charges left by missing electrons (the latter are called 'holes').

    At low temperatures, the electrons and holes in semiconductors are stuck and as immobile as valence electrons. When the temperature is high enough (room temperature is sufficiently high for many semiconductors), the thermal vibrations have enough energy to release some electrons out of their localized orbits. This is called 'thermal activation'. They then move freely through the semiconducting material. The resistance is high, however, because only a small fraction of all the valence electrons are in orbits that can be thermally activated.

    The swimming pool analogy for semiconductors is as follows: Most of the valence electrons are locked in

Electrical Current, Resistance, and Power

a completely filled pipe of water, so tilting does not produce a flow of current. A small amount of water can be thermally activated up to partly fill an empty pipe located just above the main pipe and parallel to it. This creates a water surface that can respond to a tilt or an injection of additional water. This upper pipe can 'seek its own level' and produce a horizontal flow of current.

- Shape factor

Electrical wire has a small diameter, compared with its length. The resistance of a wire grows proportionally to the length and inversely as the area of the wire. Suppose a cube of copper 1 cm on a side. The resistance value of a 1 cm cube of a material is called its 'resistivity'. The resistivity of copper is 1.7×10^{-6} Ω-cm.

Examples

1. Question: What is the resistance of a wire formed from copper having a length 7.5 cm and a cross-sectional area 0.001 cm^2?

   Answer: The resistance is

   $$\frac{1.7 \times 10^{-6} \times 7.5}{10^{-3}} = 0.013 \, \Omega$$

2. Question: You may have looked into a clear glass lightbulb and seen the filament. It is a very fine wire. For a 100 Watt bulb operating under 110 V house current, the current is $I = P/V = 100/110 = 0.91$ A. The resistance of the tungsten wire is therefore $R = V/I = 110 \text{ V}/0.91 \text{ A} = 121 \, \Omega$. Given the resistivity of incandescent tungsten is 92×10^{-6} Ω-cm, and the length of the filament is 3.5 cm, what is the cross-sectional area of the filament?

Answer: The resistance is

$$\frac{92 \times 10^{-6} \times 3.5}{A} = 121 \, \Omega$$

$$A = \frac{92 \times 10^{-6} \times 3.5}{121} = 2.7 \times 10^{-6} \, \text{cm}^2$$

- High temperature

  Raising the temperature of a metal object usually raises its electrical resistance. As the atoms vibrate more and more due to higher temperature, they move more and more from their ideal positions in the metal crystal. Vibrating atoms in the object collide with, and deflect, the electrons carrying the current. Electrons moving in response to a voltage difference in the object will accelerate, but lose or even reverse their forward momentum in each collision, then resume their acceleration toward the positive terminal. The motion is similar to that of a pinball moving downward toward the pinball player, but then hitting various obstacles which delay its progress.

  This thermal mechanism is similar to the effect of impurities and grain boundaries in a metal. In general, the obstacles slowing electron flow will be a combination of impurities, grain boundaries, and thermal vibration. Together these create a roughened path for the flow of electrons.

3.5 Ohm's Law

Section 3.3 defined electrical resistance and its unit, the Ohm. Section 3.4 explored how and why different objects and materials have different electrical resistance. This section points out an important fact about electrical conduction in resistive loads: for most conducting objects, the resistance remains constant when the applied voltage changes. Therefore, if the voltage is doubled, then the current also doubles, so their ratio, the resistance $R = V/I = (2V)/(2I)$, remains constant. Although Ohm's Law is not obeyed exactly, it is usually valid as a useful approximation.

Electrical Current, Resistance, and Power   35

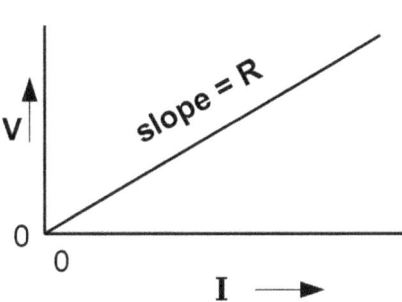

FIGURE 3.3
Current-Voltage characteristic for a Resistance obeying Ohm's Law

Figure 3.3 shows an ideal case of a material obeying Ohm's Law. The graph shows the current-voltage or I-V plot for such a material. It shows that increasing the voltage applied to a load causes a proportionately increasing current to flow. The slope of the graph is the ratio of voltage to current - i.e., the slope = R. The intercept must occur at the origin, because $I = 0$ when $V = 0$.

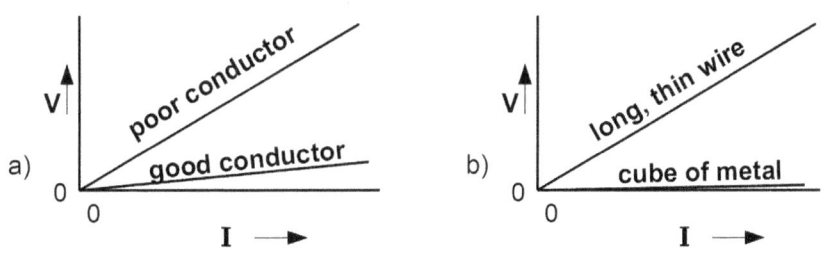

FIGURE 3.4
Examples of I-V dependence on resistance: a) larger and smaller resistances compared. b) how shape factor affects I-V plots

Figure 3.4 a) and b) show how resistance value and shape factor, respectively, affect I-V plots. In both 3.4a) and b), increasing resistance results in a curve with steeper slope. In 3.4a) the poor conductor has a higher resistance and a steeper slope. In 3.4b), a cube of metal has a very low resistance, whereas a long, thin wire has a much higher resistance, hence an I-V plot with a steeper slope.

FIGURE 3.5
I-V characteristic of a tungsten incandescent lamp

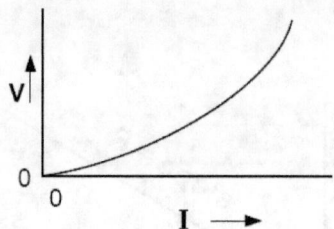

Figure 3.5 shows the I-V plot for an old-style lamp with a tungsten filament (as distinct from an LED or compact fluorescent lamp). The I-V plot is curved instead of straight, showing the tungsten lamp obeys Ohm's law rather poorly. For very small voltages, the current increases along a straight line from the origin. However, the curve bends up more and more steeply with voltage, implying a steadily higher resistance. Why does the resistance increase at higher voltage and higher current? Because the filament heats up dramatically to incandescence. More current and voltage dissipate more power in the filament. As discussed in Section 3.4, such high temperature causes tungsten atoms to vibrate violently, which deflects the electrons and impedes their average rate of flow. Hence the resistance increases.

3.6 Advanced Circuits

3.6.1 Circuits and Circulating Fluids

We have already discussed the similarity between conduction in an electrical load and water seeking its own level in your swimming pool. Hydraulic flow in a pipe is another fluid analogy which provides insight into voltage, resistance and current flow in more complicated circuits, which we are about to explore.

In hydraulics, the circuit is a set of pipes. See Figure 3.6. The flow of the fluid is analogous to the flow of charges in an electrical circuit. A paddle pump, centrifugal pump or piston pump for moving fluid in pipes draws fluid into its inlet side, and expels the fluid under pressure at its outlet. The pump and its effect on the liquid is thus very similar to a battery and its effect on charges.

Electrical Current, Resistance, and Power

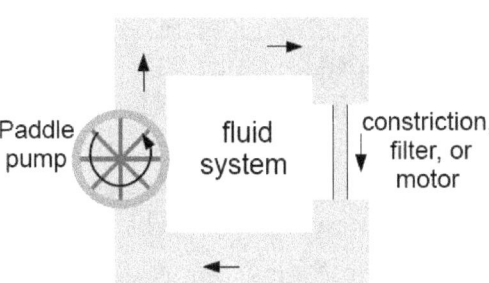

FIGURE 3.6
A typical hydraulic system which is analogous to a battery and lamp circuit

The pressure of the fluid is the analog of the voltage – i.e., the potential energy per unit charge. Finally, a typical fluid load could be a smaller diameter section in the pipe, or possibly a fluid filter for cleaning the fluid, or any such restriction to flow. Intuitively, it makes sense that a pipe having a small cross-sectional area, and/or a long length, will present a high resistance to fluid flow.

In summary, then, pressure and fluid current in a hydraulic system act like voltage and current, respectively, in electrical circuits. Logically, the flow of fluid in a pipe will be proportional to the pressure difference between the two ends of the pipe. Hence something akin to Ohm's law will apply in the hydraulic analogy.

3.6.2 Series circuits

batteries in series

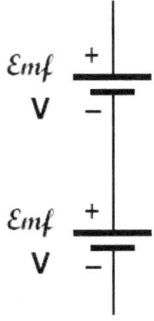

FIGURE 3.7
If two equivalent batteries are connected in series, the effect is similar to a single battery having twice the voltage.

The figure shows two batteries connected in series. 'In series' means that the positive terminal of the first battery is wired to the negative terminal of the second. The effect of this connection on potential energy is that a positive charge moving internally through the batteries has its internal energy raised by chemical action in the lower battery, and then is raised above that by the chemical action in the upper battery. The series battery produces twice the emf of a single battery.

When no current is flowing, a single battery has voltage $V = emf$, and the two batteries in series combination have $V = 2*emf$. In general, if there are multiple batteries connected in series in a circuit, the total voltage will be

$$V = emf_1 + emf_2 + ...$$

when no current is flowing. If current flows, any internal resistance of the batteries will partly reduce the output voltage from the batteries.

The mechanical analogy in Section 3.1.1 for one battery shows a man lifting boxes onto a ramp. An analogy for two batteries would be for the man to raise boxes up onto a platform. Then a second man standing on the platform could raise the boxes onto the top of a higher ramp. The result would be the potential energy of each box would be raised twice as high, and the ramp would dissipate twice the power.

resistors in series

To make a current, I, pass through two resistors connected in series, I must pass sequentially through R_1 and then R_2. There must be a potential difference across each of the resistors to force this current to flow: $V_1 = IR_1$, and $V_2 = IR_2$. If the combination of two resistors is considered as a single resistor of value R, the overall voltage is IR. Therefore,

$$\begin{aligned} V &= V_1 + V_2 & (3.4)\\ IR &= IR_1 + IR_2 \\ R &= R_1 + R_2 & (3.5) \end{aligned}$$

Electrical Current, Resistance, and Power

The result is two resistors connected in series behave as a single resistor whose value is just the sum of the two individual resistances. To find the resistance of N resistors connected in series, just find the sum of the individual resistors, $R = R_1 + R_2 + ... + R_N$.

An apt analogy is the hydraulic system. In order for fluid to pass through two constrictions in the piping, there must be a pressure difference across each. (See Figure 3.8). Hence the

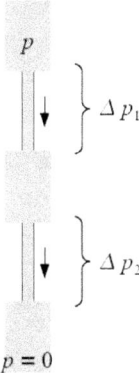

FIGURE 3.8 Pressure changes occur across the two flow-constricted parts of this hydraulic load. The pressure is nearly uniform within the wide feeder-pipe sections.

pressure at the high pressure end must be the sum of the two pressure drops:

$$p = \Delta p_1 + \Delta p_2$$

This is analogous to Equation 3.5.

Example

- Consider the case of 2 equal load resistors, as shown in Figure 3.9. Reasoning along with the hydraulic analogy, the potential difference across each load (like the pressure difference) will be $V/2$, or half the total voltage. The current is $I = V/(2R)$. The power in the top resistor (resistor number 1) becomes

$$P = I \cdot V_1 = \frac{V}{2R} \cdot \frac{V}{2} = \frac{1}{4}\frac{V^2}{R}$$

The total power in the circuit, $V^2/2R$ has been cut in half in the series circuit, relative to a circuit with just a single resistor R.

FIGURE 3.9
Equal series loads divide the voltage in half.

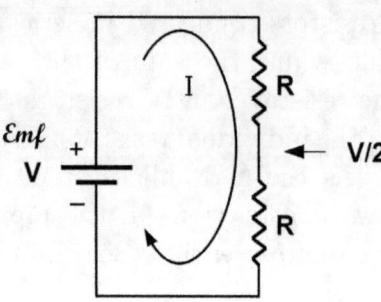

3.6.3 Parallel circuits

resistors in parallel

FIGURE 3.10
Parallel loads. The current divides, and each branch separately satisfies Ohm's Law.

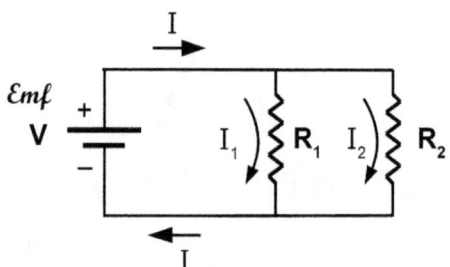

Figure 3.10 shows another common circuit configuration, two different resistive loads in parallel. Now the current flow divides between the two allowed paths. Think of a multilane freeway dividing and some of the traffic lanes going to the left, some going to the right. The total current (traffic) entering the intersection must add up to the current leaving the intersection by way of the forking paths. Therefore I supplied by the battery is equal to the sum of I_1 and I_2 :

$$I = I_1 + I_2 \qquad \text{(Kirchoff's Law)}$$

Kirchoff's Law applies to the intersection of three or more current paths. If Kirchoff's Law did not hold, charge could accumulate at the intersection. We have seen in Section 1.1 that the concentration of a small number of Coulombs at a point requires tremendous forces. Yet everyday circuits handle many Amperes

Electrical Current, Resistance, and Power

of current (= many Coulombs/second). Therefore a significant accumulation of charge is impossible and the total current entering a node must equal the total current leaving the node.

Consider that each resistor, R_1 and R_1, must separately obey the definition of resistance, Equation 3.3.

$$V = R_1 I_1, \quad I_1 = V/R_1$$
$$V = R_2 I_2, \quad I_2 = V/R_2 \qquad (3.6)$$

Let's call the effective 'total' resistance of the two resistors in parallel R_T. It follows that

$$V = R_T I_T, \quad I_T = V/R_T \qquad (3.7)$$

Substituting 3.6 and 3.7 into Kirchoff's Law,

$$\frac{V}{R_T} = \frac{V}{R_1} + \frac{V}{R_2}$$
$$\frac{1}{R_T} = \frac{1}{R_1} + \frac{1}{R_2} \qquad (3.8)$$

To find the resistance of N resistors connected in parallel, the parallel resistors formula 3.8 can be immediately extended, $1/R = 1/R_1 + 1/R_2 + ... + 1/R_N$.

Example

- Consider the case of 2 equal load resistors in parallel, as shown in Figure 3.11. The current will be double what it is through a single load with the same voltage. Hence the effective resistance is $R/2$. Hence the total resistance has been cut in half by adding the additional load in parallel. This is true in general for parallel resistors: adding any additional resistance in parallel decreases the overall resistance because it always allows more current to flow for the same voltage. The current in each resistor individually is the same as it was in the single load case, and the voltage is of course the same for parallel loads as for single load. So for any R_1 and R_2 in parallel, R_T will always turn out to be smaller than either R_1 or R_2 separately.

The power in the left resistor becomes

$$P = I \cdot V = \frac{V}{R} \cdot V = \frac{V^2}{R}$$

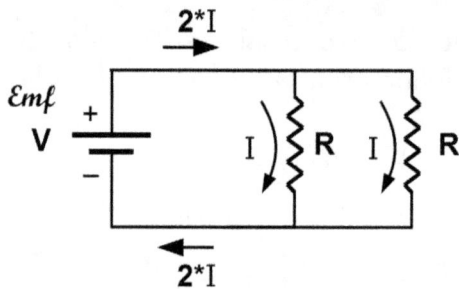

FIGURE 3.11 Equal parallel loads double the total current in the battery circuit.

The total power in the circuit, $2 * V^2/R$, has been doubled in the parallel circuit, relative to a circuit with just a single resistor R.

batteries in parallel

Two batteries may be connected in parallel. 'In parallel' means that the positive terminal of the first battery is wired to the positive terminal of the second, and similarly, negative is connected to negative. The parallel battery connection produces the same emf as a single battery. The advantage is the combined battery can maintain this emf while delivering a larger current than a single battery. Or, a given current can be delivered for a longer time before the chemical energy of the two batteries in parallel is depleted.

One way of viewing the current-delivering capability of batteries connected in parallel is to look at the effective internal resistance of the combined connection. The internal resistances of multiple batteries add in parallel. Hence the internal resistance of N parallel-connected batteries is N-times smaller than that of one battery.

Parallel Battery Example

1. Question: A single 1.5 V battery is connected to a short loop of heavy wire. (See Figure 3.12a) The wire has a resistance of 10 milliohm. If the internal resistance of the battery is 0.3 Ω, how much current flows in the circuit?

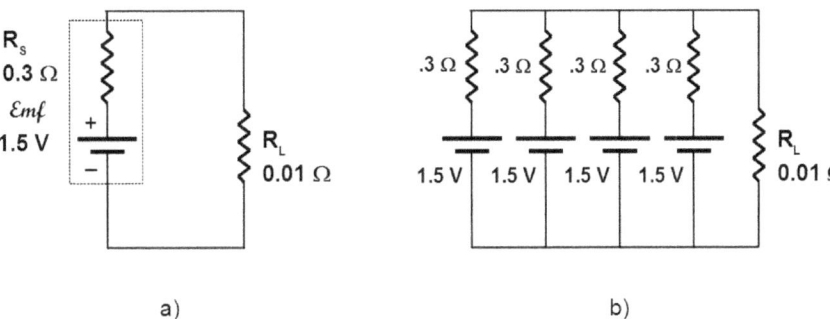

FIGURE 3.12
a) The internal resistance R_S of a battery is effectively in series with the load R_L.
b) Four batteries in parallel.

Answer: The circuit acts as if an ideal battery (no internal resistance) is connected to a .3 Ω resistor in series with a 0.01 Ω resistor. The current flow is therefore

$$I = \frac{1.5}{0.3 + .01} = 4.85 \text{ A} \tag{3.9}$$

2. Question: Now four of the same 1.5 V batteries in parallel are connected to the same short loop of heavy wire, whose resistance is still 10 milliohm. (See Figure 3.12b) If the internal resistance of each battery is still 0.3 Ω, how much current flows in the circuit?

Answer: Again each battery has a 0.3 Ω resistor in series with it. One end of each resistor is held at 1.5 V by its corresponding ideal battery. The other end of all four internal resistors is a node in the circuit. Therefore the internal resistors all have identical voltage difference applied to them.

Since they all have the same resistance value, they must all have the same current flowing. Calling that current I,

$$\begin{aligned} EMF &= IR_{int} + (4\,I) * R_L \\ 1.5 &= I * 0.3 + 4 * I * 0.01 \text{ A} \\ I &= 19.6 \text{ A} \end{aligned}$$

The above calculation is the same as replacing the four parallel batteries with a new battery that has the same emf, but a new effective internal resistance. The new internal resistance will be equal to the four internal resistors in parallel.

3.6.4 More complex series-parallel circuits

An infinite variety of series-parallel resistor circuits is possible. Many are subject to a simple analysis, as follows:

1. Focus attention on subunits of the circuit that are simple series and simple parallel configurations. The simple series subunit must have the same current flowing through both resistors that are in series. The simple parallel subunit has a node where the current divides into two or more branches, and a downstream node where all the branches re-unite.

2. Find the total resistance of each series or parallel subunit of the circuit, using 3.5 or 3.8, resp. Re-draw the circuit with that respective series or parallel subunit replaced by a single resistor having the correct total resistance.

3. Repeat steps 1) and 2) until the complex series-parallel combination is reduced to a single, grand total resistance. Using this grand total resistance and the battery voltage for the circuit, find the total current through the battery.

4. Now use the above information about voltage and current to 'drill down' to one of the subunits. If this is a series element, the battery current passing through this element will cause a 'voltage drop'. Therefore the rest of the circuit after the series element must have the same battery current and the divided voltage left after subtracting the voltage drop from the battery *emf*. If instead we drill down to a parallel subunit, each branch receives the full battery *emf*. The ratio of resistances of the branches determines the fraction of the total current entering into each branch.

5. Iterate the drilling down process until it assigns a voltage and current value to every parallel and every series resistor.

Example

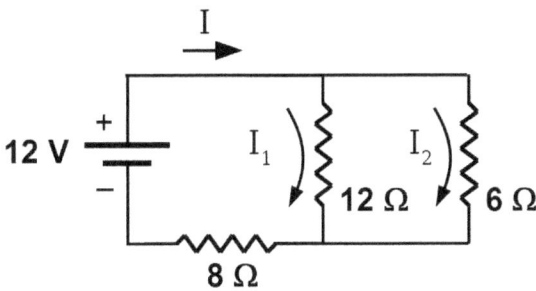

FIGURE 3.13
Three different resistances in series-parralel.

- Consider the 3 resistors in series-parallel shown in Figure 3.13. There is a resistor in series with a parallel combination. First find the total resistance of the parallel combination: $1/R_T = 1/12 + 1/6 = 1/4$ so $R_T = 4\ \Omega$. Replace this parallel combination with the single value $4\ \Omega$. This is in series with the $8\ \Omega$ resistor, so the total of resistance in the circuit is $12\ \Omega$. Since the total *emf* is $12 Volts$ and total circuit resistance is $12\ \Omega$, the current through the battery is 1 A. This current must pass through both elements in series - so the voltage drop across the $8\ \Omega$ resistor is 8 V. This leaves 4 V across the parallel combination. Since both $12\ \Omega$

and 6 Ω experience 4 V, the current in the 12 Ω resistor is 0.33 A and the current in the 6 Ω resistor is 0.67 A. Now that all currents and voltages throughout the circuit are revealed, it would be easy to calculate power in each resistor. This completes the analysis.

Chapter 4

The Electric Field

Objective: The next two chapters introduce electric and magnetic <u>fields</u>. These are force fields that allow transfer of force across empty space, from one object to another, without the two objects actually touching. The electric field transfers force from one charge to another. The magnetic field transfers force from one current to another. We can think of the field as always surrounding a charge or a current, even though there may actually be no other charge or current at the receiving end to experience the electric or magnetic force. We will also see that the fields contain energy. In Chapter 8.3 we will learn that the fields also are the mechanism by which energy travels over great distances at the speed of light.

In Chapter 1, Equation (1.1) described the force between two charges. This force can be broken down into two parts: (1) the part that one of the charges, say Q_1 causes, and (2) the part that the second charge, Q_2, causes. The part caused by Q_1 is called the *electric field*. The electric field from a single charge is a vector field emanating radially from it in all directions. Q_2 experiences a force due to being immersed in this electric field.

4.1 Defining The Electric Field

We can rearrange Equation (1.1) into two factors as follows:

$$F = k\frac{Q_1 Q_2}{R^2}$$

$$F = \left(k\frac{Q_1}{R^2}\right) Q_2$$

The grouping of terms in the above equation suggests the following equations:

$$E = k\frac{Q_1}{R^2} \quad \text{and} \quad (4.1)$$

$$\vec{F} = \vec{E} Q_2 \quad (4.2)$$

Equations (4.1) and (4.2) define the electric field surrounding a point charge Q_1, and the force on a test charge Q_2 in that field, resp. In Equation (4.1), E depends only on Q_1 and the distance from charge Q_1. The electric field due to Q_1 exists in all space surrounding Q_1. It has the units Newtons/Coulomb. For example, if the electric field at some point in space is 10 N/C, and a charge q = 1 C is placed there, then the force on the 1 C charge will be $F = E * q = 10 * 1 = 10N$. If instead a charge q = 2 C were placed there, then the force on the 2 C charge would become 20 N. Hence the electric field is the force per coulomb for whatever charge is placed in that electric field.

The direction of the electric field is the same as the direction of the force on a positive charge placed in the field. Hence the electric field can be depicted as a pattern of arrows representing vectors, filling all space. The arrow at a given place has length proportional to the force on a 1 C charge placed at that point, and direction the same as the direction of the force. The electric field surrounding a positive and a negative charge is shown in Figure 4.1a and b, resp. The E-field arrows in Figure 4.1a point outward from Q, because any positive charge will be repelled by Q. The E-field arrows in Figure 4.1b point towards charge -Q, because any positive charge will be attracted toward -Q.

The Electric Field

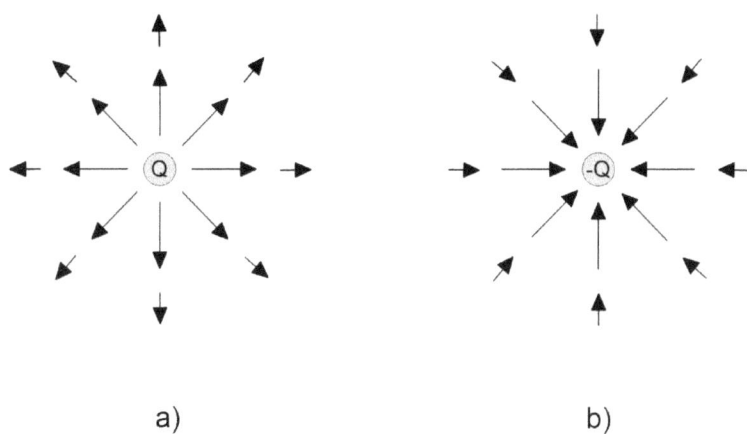

FIGURE 4.1
Arrows representing the electric field in the vicinity of (a) positive (b) negative charges.

4.1.1 Uniform Electric Field

The magnitude of the electric field due to a point charge, discussed in Section 4.1, obeys the inverse square law dependence upon distance. The direction of the E-field due to a point charge depends on which side of the point charge we are located. To the right of a positive charge, the E-field points to the right. To the left of a positive charge, the E-field points to the left. *Et cetera*. This section considers a uniform electric field — that is, one which is constant in magnitude and direction throughout a given region of space.

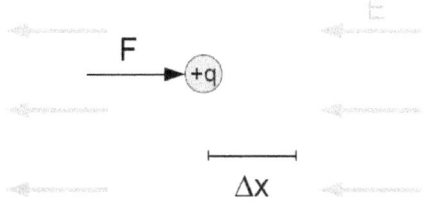

FIGURE 4.2
Force F displacing a charge +q a distance Δx in a uniform E-field

Voltage in a uniform E-field

Suppose there is a uniform electric field $\vec{E}$ pointing toward the left, as shown in Figure 4.2. According to Equation (4.2), a positive test charge q placed in this field experiences a force qE. If we apply force F, equal and opposite to the electric force, to hold q stationary, there is no work done. But if F moves q a distance Δx to the right, it performs work on q,

$$\Delta W = \Delta PE = qE\Delta x \tag{4.3}$$

This is analogous to lifting a mass m against the gravitational field and doing work mgh. The potential energy of q has increased by $\Delta U = qE\Delta x$. According to Equation (1.4), the change in potential energy per unit charge is equal to the difference in voltage between the two locations, that is,

$$\Delta V = \frac{\Delta U}{q} = |E|\,\Delta x \tag{4.4}$$

Hence the magnitude of E-field is

$$|E| = \frac{\Delta V}{\Delta x} \tag{4.5}$$

Now consider the direction of the E-field. The E-field points in the negative x-direction. We have increased the energy, and hence the voltage, by pushing q in the positive x-direction, and *opposite* the direction of the E-field. So E-field is given by the negative slope of the voltage:

$$\vec{E} = -\frac{\Delta V}{\Delta x} \tag{4.6}$$

The direction opposite to the E-field vector is the direction of most rapid increase in voltage. Changes in position perpendicular to the E-field produce no change in voltage. The latter case is similar to the motion of a mass horizontally producing no change in the gravitational potential energy because the motion is perpendicular to the direction of the force field. The force has zero component in the direction of motion.

Important Voltage Principle

- The voltage *increases* as we approach a + charge or concentration of + charges. The voltage *decreases* as we approach a − charge or concentration of − charges.

The Electric Field

E-field and Voltage Examples

1. Question: The voltage difference between the plates of a parallel-plate capacitor is 15 V, and the distance between the plates is 0.1 mm. What is the electric field between the plates?

   Answer: According to Equations (4.6)-4.5, the E-field is the negative slope of the voltage. The E-field points towards the lower voltage plate, and the magnitude of E is

   $$\Delta V/\Delta x = 15\text{V}/(1 \times 10^{-4}\text{m}) = 150{,}000 \text{ V/m}$$

2. Question: The *breakdown* E-field is the E-field just large enough to cause an electric arc to jump through an insulator. In dry air the breakdown field is 3×10^6 V/m. If the air-gap between two metal plates is 2.54 cm, what is the breakdown voltage?

   Answer: The magnitude of E is $\Delta V/.0254$ m $= 3 \times 10^6$ V/m, so $\Delta V = 7.6 \times 10^4$ V

4.2 Lines of Force

A little bit different way of depicting the electric field is shown in Figure 4.3 Here the E-field is depicted as lines of force emanating from the point charge Q. The direction of the lines of course shows the direction of the field. The *density* of the lines is intended to give the *magnitude* of the E-field. Note because this drawing is 2-dimensional, it does not do justice to the 3-D picture of the lines of force. The electric field should grow as $1/R^2$ for a point charge, whereas the density of lines in a 2-D drawing such as Figure 4.3 appears to vary as $1/R$. The figure *does* show the lines correctly in the sense they are evenly distributed around the charge. If the drawing were done in 3-D instead of 2-D, the lines would still emanate evenly distributed in all directions from the charge. In a 3-D drawing, however, the density would decrease more quickly, as $1/R^2$. This would be the correct R-dependence.

FIGURE 4.3
Lines of force representing the electric field in the vicinity of positive charge Q.

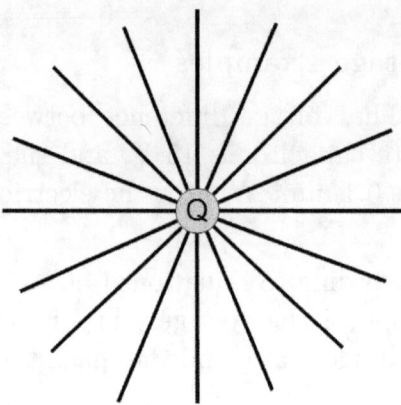

- Question: Why does a point charge Q with lines of force evenly distributed in all directions indicate an E-field with $1/R^2$ dependence on distance? What is the number of lines of force?

Answer: The key is that the area of a sphere equals $A = 4\pi R^2$. Suppose a number N lines of force emanate from point charge Q. At radius R_1, there are N lines of force emanating from an area $A_1 = 4\pi R_1^2$. Therefore the density of lines of force is $N/A_1 = N/(4\pi R_1^2)$. The density of lines must be equal to E-field, so both

$$E(R_1) = \frac{N}{4\pi R_1^2} \quad \text{and}$$

$$E(R_1) = k\frac{Q}{R_1^2}$$

Equating the two righthand sides of the above equations,

$$\frac{N}{4\pi R_1^2} = k\frac{Q}{R_1^2} \quad \text{so}$$

$$N = 4\pi kQ \tag{4.7}$$

So the number of lines coming from a charge Q is $N = 4\pi kQ$. Now repeat the above calculation of the number of lines N at a larger radius R_2. At R_2, there will be the same number N lines of force coming out of a larger area $A_2 = 4\pi R_2^2$. So the density of lines of force will be smaller, $N/A_2 = N/(4\pi R_2^2)$. This correctly indicates the smaller size of the E-field at R_2.

4.3 Rules for E-lines of Force

The rules for E-lines of force are as follows:

1. The force on a test charge Q_2 is parallel to the line of force passing through Q_2. The magnitude of the force at Q_2 is proportional to the density of the lines of force at Q_2.

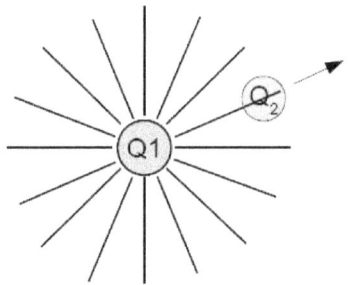

FIGURE 4.4
Arrow indicates direction of force on test charge Q_2.

2. Lines of Force start on + charges and end on − charges. They cannot cross, start or stop, except on charges. For example, if lines of force could cross, then we could put a small test charge on the crossing point, and the lines would indicate two conflicting directions for the force on the test charge. As a further example, consider the drawing of 4.5, showing lines of force going between equal and opposite + and − charges. The pair of equal and opposite charges is called a *dipole*, and the result is a *dipole field*.

    Every line of force that emanates from the + charge eventually must stop on the − charge. Of course, the lines going left from the + charge must travel very far in order to come back and land on the − charge from the right.

3. Lines of force cannot penetrate a metal object. This rule assumes a static situation. If E-fields penetrated a metal, this

FIGURE 4.5
Representation of a dipolar electric field.

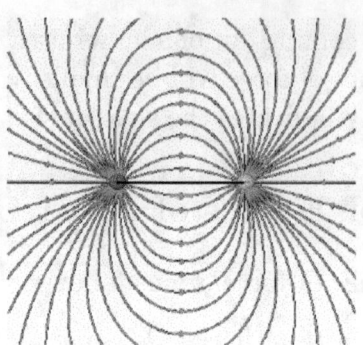

would propel electrons and create a very large current. The current would transfer charge in the metal object and cancel the applied field. Assuming equilibrium has been reached, and there are no generators supplying steady current, the electric field must be zero inside a metal.

4. Lines approach a metal surface at right angles. This follows from the previous rule. If lines of force approached the surface at an oblique angle, the E-field would have a vector component parallel to the surface of the metal. The same arguments as above rule out any E-field component in an equilibrium situation with no currents in the metal.

5. For a large, uniformly charged flat plate, the field is *uniform* — meaning the lines are evenly spaced and straight — and perpendicular to the plate. The argument for straight, parallel E-lines is based on symmetry. The force on a test charge placed near the plate must be normal to the plate, because there is no reason for any particular preferred direction if the force had any parallel component. The lines must be evenly spaced coming from or going to the plane, because the lines terminate on charges and the charges are evenly spaced over the plane. A single plate is shown in 4.6a, and a pair of parallel plates with equal and opposite charges is shown in 4.6b.

The Electric Field

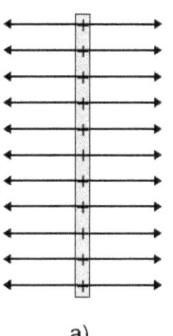 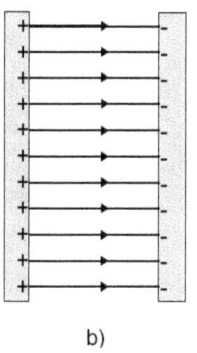

FIGURE 4.6
Lines of force (a) surround a uniformly charged flat plate and (b) stretch between two parallel plates.

4.4 Parallel plate Capacitor

If the plates are made of metal, and hold equal and opposite charges, the E-field and E-lines are similar to the uniformly charged parallel sheets of Figure 4.6b. This is called a parallel plate *capacitor*, and is very important for storing charge. Charges are free to move on a metal plate, but they will be evenly spread over most of the plate. Any excess density of charge in one spot would cause field lines to emanate with some component parallel to the metal, and this would violate the above rules. Rather, charges repel and spread very evenly across most of the plate. A slightly higher concentration occurs near the edges of the plate. If the plates are large compared to the capacitor plate separation, this small additional concentration near the edge is not important.

Figure 4.7 shows a pair of parallel metal plates holding equal and opposite charges, and arrows representing the E-field between the plates. By Rule 2 the E-lines of force start on the + charges and terminate on the − charges. Very few E-lines reach outside the capacitor plates; the E-field is practically zero outside the capacitor.

When the + and − terminals of a battery are connected to the two plates of a capacitor, electrons flow out of the (-) terminal onto the negative plate, and electrons flow off the positive plate to the (+) terminal of the battery, completing the battery circuit and charging the capacitor. The E-field between the plates grows while the battery is charging the plates. According to Equation

FIGURE 4.7 Electric field in a parallel plate capacitor. The capacitor has metal plates. The plates have equal and opposite charges on them.

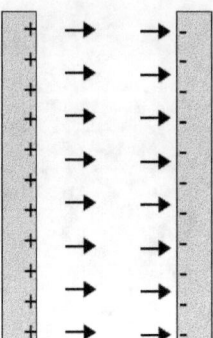

(4.5), the slope of the voltage grows, and therefore the voltage difference between the plates also increases. The final voltage between the plates will be determined by the battery voltage. If the battery *emf* is V and plates are separated by distance d, the fully-charged capacitor will have electric field $E = V/d$.

How much charge does the capacitor hold when it is charged up to voltage V? It seems logical that the charge should depend on the E-field (and therefore V) as well as the plate area and plate separation. We introduce Gauss's Law in the following section to find out the relationship between the amount of stored charge and the E-field in a capacitor.

4.5 Gauss's Law

We have seen from equations (4.1) that the E-field around a point charge Q is $E = kQ/R^2$, and this equation gives the force on any size test charge q anywhere around charge Q, through the relationship, $F = qE$. Point charges are a common charge geometry. Another very useful geometry is a uniform field. What is the relationship between the amount of charge stored on a set of capacitor plates and the strength of the E-field between the plates? The mathematician Gauss saw the connection between these different situations is the number of lines of force. Gauss's Law is as follows:

- Consider a collection of charges surrounded by an imaginary, closed container. Then the number of lines of force emanating from the container is equal to

$$N = 4\pi k Q_{total} \qquad \text{Gauss's Law} \qquad (4.8)$$

where Q_{total} is the total net charge inside the container.

If there are both positive and negative charges in the container, then Q_{total} is the total *net* charge— i.e., the positives minus the negatives. For example, if the container has 5 + charges and 2 − charges, the net is 3 + charges. Similarly, if some E-lines leave the container and some point back into it, Gauss's law uses the *net* of number leaving minus the number entering.

4.5.1 Gauss's Law and Charged Spheres

It is very useful to apply Gauss's law to a spherical distribution of charge. As we explore in the example below, Gauss's Law leads to the following general principle:

> Outside any spherically symmetric configuration of charge, the electric field— and forces due to it— act as if all the charge were located at the **center** of the sphere.

Consider a metal sphere of radius a with a positive charge Q. Recall from Section 2.2.2 that Q distributes itself uniformly over the sphere to minimize potential energy. Figure 4.8 shows the configuration.

According to Gauss's Law Equation (4.8), the number of Lines of Force penetrating the imaginary sphere is $N = 4\pi k Q$. Equate this to the flux of E-field flowing out of the sphere of radius R: $N = E * A = E * 4\pi R^2$. Then solving for the E-field at radius R, $E(R) = kQ/R^2$. This is the same E-field obtained for a point charge Q, which is very interesting: outside the sphere, the E-field is the same as if the entire charge Q were located at the center of the sphere. This is similar to the fact that the Earth's gravitational attraction pulls towards the Earth's center with a

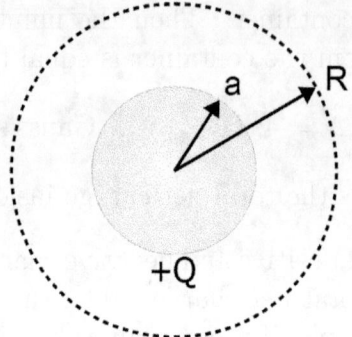

FIGURE 4.8 Metal sphere of radius a has charge Q. An imaginary surface is drawn around the sphere at radius R.

force that would be the same if all the mass of the Earth were located at the center.

Now let's put the imaginary surface *inside* the metal sphere. That is, $R < a$. In this case, the imaginary surface encloses no charge. By Gauss's Law, the E-flux through the surface of the imaginary sphere must be zero. By symmetry, if there were any E-field inside the sphere, it must point in the radial direction. The interesting conclusion is that there is no E-field inside a hollow metal sphere, even if there is a very large charge Q on the surface of the sphere. This result has a practical application: If you enclose a delicate instrument inside a metal container, and raise the container to a very high voltage by charging it up, there is no electrical effect on the delicate instrument. This is known as electrostatic shielding, and is commonly used to shield sensitive circuits such as smartphones and computers from the electrical disturbances and noise that is all around us.

N was easy to count for a point charge and a spherical surface when we just counted up the E-field magnitude at the surface multiplied by the area of the surface. For more complicated surfaces, the E-field may not be the same over the whole surface, and may not be directly pointing out of, or *normal* to the surface at some point. For example, in Section 4.5.2 we analyze the parallel plate capacitor. Gauss's law will require adding each part of the surface multiplied by the E-field strength at that point, and considering only the component of E that actually points normal to the surface.

4.5.2 Parallel Plate Capacitor E-field

We now apply Gauss's Law, Equation (4.8), to calculate the E-field due to the charge stored in a parallel plate capacitor, Figure 4.7. Figure 4.9 shows the capacitor embedded in plate extensions. The extensions rule out any non-uniform concentration of charge or bulging of the E-field at the ends of the plates. Therefore the charge is distributed perfectly evenly within the capacitor and the E-field is perfectly parallel and normal to the metal plates. In a real capacitor, the area is usually much bigger than the plate separation, so end effects are usually negligible anyway. Let the area of each capacitor plate be A, and the distance between the plates be d.

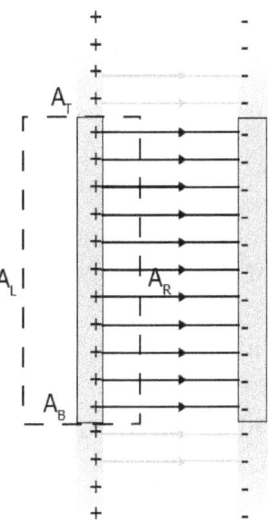

FIGURE 4.9
A parallel plate capacitor embedded in plate extensions. We apply Gauss's law to the charge on the left capacitor plate and the E-field it originates.

The dotted rectangle in the figure indicates an imaginary closed surface surrounding the left capacitor plate. This surface has left and right 'caps' which are parallel to, and equal in area to, the plates of the capacitor. The top and bottom areas, together with front and back areas which are not visible but are parallel to the page in front and behind the page, complete the imaginary surface. The surface entirely encloses the left capacitor plate.

Applying Gauss's Law requires calculating the flux of E-field leaving the enclosed surface. The E-field leaves the surface only through the right side, A_R. This contribution to the emanating flux is $E * A_R = EA$. There is no E-field leaving or entering through A_L, because E is zero outside the capacitor. That is because all the E-lines start on + charges and terminate on the − charges within the capacitor, so the field is entirely inside the capacitor. E is not zero along A_T and A_B, the top and bottom surface. However, there is no component of E *normal* to these surfaces, and so no E-lines cross in or out of the enclosure at these places. Similarly, there is no component of E-field crossing the front and back surfaces, which complete the enclosure above and below the page, and which are not shown in Figure 4.9.

Gauss's Law calculation of Equation (4.8) is therefore $EA = 4\pi kQ$, so E inside the parallel plate capacitor is finally,

$$E = \frac{4\pi k}{A} Q \qquad (4.9)$$

4.5.3 Charge-voltage relationship for Capacitors

We want to calculate how much charge Q a capacitor stores when voltage difference V is applied between the capacitor plates. Assume again capacitor plate area A and plate separation d. The electric field E is constant inside the capacitor. The E-field is the negative slope of the voltage. Using the magnitude of E, from Equation (4.5) and substituting the Gauss's Law result for E, Equation (4.9),

$$|E| = \frac{\Delta V}{d} = \frac{4\pi k}{A} Q \qquad \text{, then solving for Q,}$$

$$Q = \left(\frac{1}{4\pi k} \frac{A}{d}\right) \Delta V \qquad (4.10)$$

The expression in parentheses contains only constants and the geometric quantities — area and plate separation. We define this as the *capacitance* C:

$$C = \frac{1}{4\pi k} \frac{A}{d} \qquad \text{Parallel plate capacitance formula} \qquad (4.11)$$

The Electric Field

Assume the negative plate is at voltage 0 Volts, and the positive plate is at voltage V. Then $\Delta V = V - 0 = V$. Equation (4.10) becomes

$$Q = CV \quad \text{Charge-voltage relationship for capacitor C} \quad (4.12)$$

There are capacitors with other geometrical configurations besides parallel plates. Regardless of the geometry, the capacitance is defined, in general, by Equation (4.12) as the ratio between the charge stored and voltage applied.

Capacitance and Charge storage Examples

1. Question: A capacitor consists of a pair of metal plates. The area of each is 10 m² and the distance between the plates is 0.2 mm. What is the capacitance in Farads? in *microfarads*? in *picofarads*?

   Answer: The capacitance formula gives $C = 10/(4\pi k * 0.0002) = 4.42 \times 10^{-7}$ F $= 0.442~\mu$F $= 442,000$ pF.

2. Question: The voltage difference between the plates of a parallel-plate capacitor is 15 V, and the capacitance is 0.1 μF. What is the charge stored in the capacitor?

   Answer: The capacitor charge-voltage relationship gives $Q = CV = 0.1 * 15 = 1.5 \times 10^{-6}$ C

3. Question: The high voltage sphere of a van der Graaf machine achieves a voltage of 75,000 V. If the capacitance between the sphere and the ground is 3.3 pF, how much charge is stored on the vdG's sphere?

   Answer: $Q = CV = 3.3 \times 10^{-12} * 7.5 \times 10^4 = 248 \times 10^{-9}$ C.

4.6 Practical Capacitors

We have discussed the parallel-plate capacitor in terms of an interesting example of E-fields and the rules that govern them. Now the story turns to some important practical uses of capacitors. Two useful things that capacitors do in circuits is (1) store energy; and (2) provide timing.

4.6.1 Energy Storage in a Capacitive Circuit

There are two ways to look at energy storage in capacitors—by analyzing circuits containing capacitors or by analyzing their electric fields. Here we look at the energy flow when a capacitor becomes charged in a circuit. When we discuss electromagnetic waves in Chapter 8, we will see the same result comes from looking at how much mechanical energy it takes to create E-field in a capacitor.

Consider a capacitance C charged to a voltage V with a charge $Q = CV$, as per Equation (4.12). The capacitor has a separation of charges $+Q$ and $-Q$, so it should be possible to obtain energy from the capacitor by allowing the charge from one plate to travel through a circuit to reach the opposite sign of charge on the other plate. Such a circuit is shown in Figure 4.10. One might jump to the conclusion that the available energy is the charge multiplied by the potential difference between the plates, that is Q*V (which has the units of Joules). However, this is not correct. The first small increment of charge travels through the load with potential change V, but this of course decreases the voltage between the plates so that the next increment of charge has less potential energy to contribute.

FIGURE 4.10
Discharge of a capacitor C through a load.

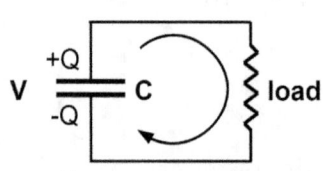
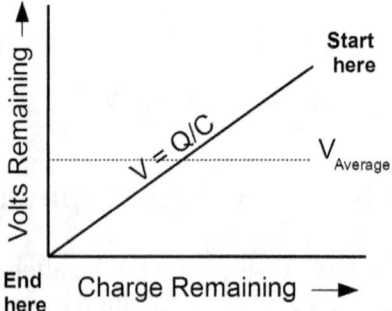

The Electric Field

Therefore imagine the load is adjusted to allow a steady flow of charge, so the voltage and the charge decrease steadily from maximum to zero. The dashed line in Figure 4.10 indicates the voltage V decreasing proportionally as the charge Q decreases. Clearly if V_0 were the initial voltage, then the charge on the capacitor drops through an *average* voltage $V_0/2$. So the energy available from a capacitor charged to voltage V with charge Q is

$$U = Q\frac{V}{2} \quad \text{and using } Q = CV,$$
$$= \frac{1}{2}CV^2 \qquad (4.13)$$

Capacitor Energy Example

- Question: A 0.22 F capacitor is charged to a voltage of 6V. How much energy is stored?

  Answer: Applying Equation (4.13), the energy is $\frac{1}{2}$ 0.22 F*(6 V)^2 = 4.0 J.

4.6.2 Capacitor Dielectrics

The examples in 4.5.3 showed it is not possible to store much charge in a parallel plate capacitor of reasonable dimensions. The capacitance is proportional to the plate area, and can also increase if the plate separation is decreased. However, because the breakdown field for (dry) air is $\approx 3 \times 10^6$ V/m, there is a limit to how thin the space between the plates can be.

Materials inserted between the plates, called *dielectrics*, are used to multiply the capacitance many times what it would be if filled only with air. Figure 4.11 shows a parallel plate capacitor with a dielectric slab in between the plates. The positive capacitor plate attracts some negative charge to the surface of the dielectric near it. Likewise, the negative capacitor plate attracts the same amount of + charge. This reduces the average field between the plates, because some E-field lines from the positive plate terminate and do not penetrate the dielectric. The E-field is smaller in

FIGURE 4.11
Dielectric placed between the plates of a capacitor reduces the average E-field.

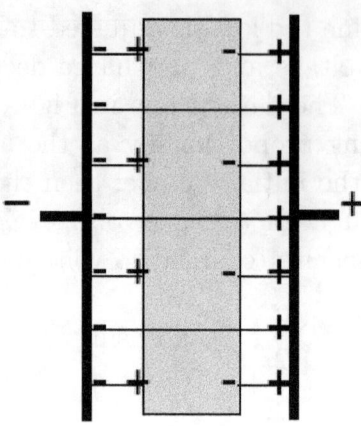

Material	Dielectric Constant, ϵ	Breakdown E-field, V/m
air	1	3×10^6
pure water	80	30×10^6
tantalum oxide	20	200×10^6
aluminum oxide	10	13.4×10^6
mylar	3.2	100×10^6
polystyrene	2.6	19.7×10^6

TABLE 4.1: Dielectric constant and Breakdown field of common dielectric materials

the slab, relative to the E-field if the slab were not there. Because the E-field determines the slope of the voltage, the voltage difference between the capacitor plates is now less with the dielectric in place. So the capacitor is now holding the same charge with a smaller voltage. This means the capacitance is bigger, because C = Q/V, and V is in the denominator.

Table 4.1 lists the *dielectric constants* of some commonly used materials. Although water is not used to make capacitors, the dielectric constant is shown for reference. When a dielectric material fills the entire space between the plates of a capacitor, the capacitance is multiplied by the dielectric constant, ϵ, of that ma-

The Electric Field

terial. Hence

$$C = \frac{\epsilon}{4\pi k}\frac{A}{d} \qquad \text{Parallel plate capacitance with dielectric} \quad (4.14)$$

Another advantage of using dielectrics is they typically improve the breakdown field in the capacitor. Table 4.1 also shows the breakdown fields for dielectric materials. A material such as mylar, familiar to us as a metalized balloon material, has great dielectric strength. Because the breakdown field is so large, very thin layers of mylar are sufficient to hold back the voltage applied to a capacitor. Therefore, according to (4.14), both the plate separation, d, and the plate area, A, can be scaled down proportionally to make a smaller-sized capacitor with the same value of C.

4.6.3 Timing In a Resistor-Capacitor Circuit

In Figure 4.12 a capacitor starts with charge Q_0 and voltage V. Then the switch is closed, allowing the charge to circulate through load resistor R. The charge on the capacitor drains quickly at first. But when some charge has left the capacitor, the voltage, V = Q/C, has decreased. The reduced voltage drives a proportionally smaller current through R. Thus the voltage and the charge on the capacitor approach zero asymptotically. The actual time-dependence is an *exponential decay*.

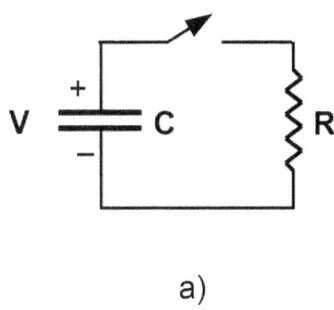

a)

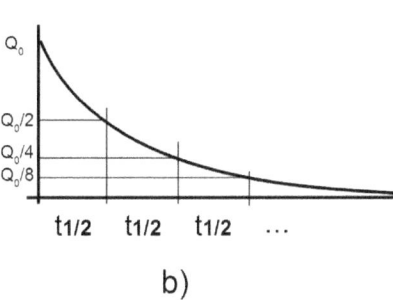

b)

FIGURE 4.12 a) Circuit consisting of capacitor, resistor, and a switch. b) When the switch is closed, the initial charge decays — rapidly at first.

An exponential decay characterizes many physical processes, including radioactive decay and chemical decomposition reactions. In each of these processes, as in RC decay, the rate of change of a reactant is proportional to the amount of reactant itself remaining. You may be familiar with the terminology, *half-life*, which describes radioactivity. For example, ^{131}I decays with a half-life of 8.02 days, and is an immediate concern in case of a nuclear power plant breach. On the other hand, ^{99}Tc decays with a half-life of 210,000 years, and is a worrisome component of nuclear waste from normally working reactors.

4.7 Capacitors in Series and Parallel

Figure 4.13 shows two capacitors in parallel. The question is what is the combined capacitance in this configuration and in the series configuration of Figure 4.14? Suppose the parallel capac-

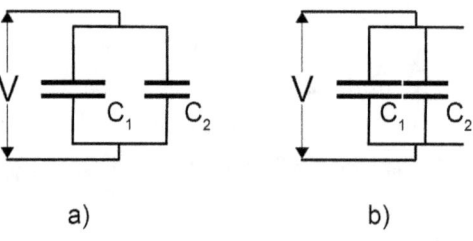

FIGURE 4.13 a) Parallel capacitor connection. b) If the capacitors had the same plate separation, they could slide together to provide a simple answer for the total capacitance.

itors happen to have the same plate separation, d. Then create a bigger capacitor by sliding the two capacitors so the plates are contiguous, as in Figure 4.13b). The combined capacitor has area

The Electric Field

$A_1 + A_2$, so the total capacitance is

$$\begin{aligned} C_T &= \frac{1}{4\pi k} \frac{A_1 + A_2}{d} \\ &= \frac{1}{4\pi k} \frac{A_1}{d} + \frac{1}{4\pi k} \frac{A_2}{d} \\ &= C_1 + C_2 \quad \text{capacitors in parallel} \end{aligned} \quad (4.15)$$

The result (4.15) is still correct when parallel capacitors have different plate separations. In the general case of parallel-connected capacitors, C_1, C_2, and C_T all have the same voltage applied, and Equation (4.15) is therefore just echoes that the total charge is the sum of two parts— i.e., $Q_T = Q_1 + Q_2$.

For series capacitors, the voltage divides between the two capacitors. This reduces the amount of voltage on each capacitor,

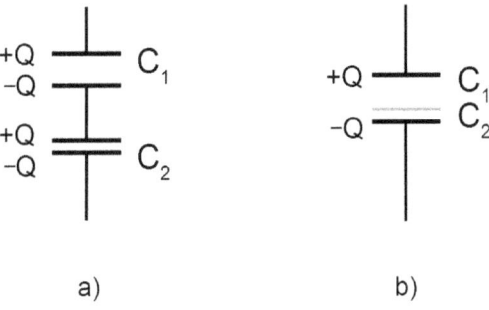

FIGURE 4.14
a) Series capacitor connection.
b) If the capacitors had the same plate area, they could slide together to provide a simple answer for the total capacitance.

which reduces the charge and hence the capacitance of the series combination. Suppose the capacitors have the same plate area, A. Now slide the upper capacitor in Figure 4.14 down to the lower capacitor until the two merge, as in Figure 4.14b. It is important to realize that the lower plate of the upper capacitor and the upper plate of the lower capacitor started with no net charge before the voltage was applied; and therefore the charges were equal and opposite with voltage applied; and finally the charges cancel when the two plates merge.

The result in Figure 4.14b is a C_T having the same area as the two starting capacitors, but a different plate-separation, $d_T = d_1 + d_2$. Therefore

$$\begin{aligned}\frac{1}{C_T} &= \frac{4\pi k}{A}(d_1 + d_2) \\ &= \frac{4\pi k}{A}d_1 + \frac{4\pi k}{A}d_2 \\ &= \frac{1}{C_1} + \frac{1}{C_2} \quad \text{capacitors in series} \end{aligned} \quad (4.16)$$

Equation (4.16) is still correct when series capacitors have different plate separations. In the general case of series-connected capacitors, C_1, C_2, and C_T all have the same charge. Therefore Equation (4.16) effectively states that the total voltage is the sum of two parts— i.e., $V_T = V_1 + V_2$.

Chapter 5

The Magnetic Field

Objective: Like the electric field, the magnetic field can extend through materials and across empty space to exert forces. You will find that the magnetic field is somewhat more complicated geometrically. Although the electric field emanates from static charges, only moving charges— e.g., electric currents— can generate a magnetic field. Another very important difference is that magnetic forces are potentially much stronger than electrical forces. Magnetism makes possible very strong, high speed, compact, and efficient motors. The ingredient leading to stronger forces is that certain materials, called ferromagnetic materials, dramatically multiply the magnetic fields produced by electric currents.

Like our introduction to the electric field in Section 4.1, this section will write the magnetic force between two currents, and then break the force equation into two parts. This reveals an intermediary magnetic force field, B, emanating from one current and carrying the force to the other current. The magnetic force between two parallel wires of length L, carrying currents I_1 and I_2 in the same direction, and separated by a distance R, is

$$F = \frac{\mu_0}{2\pi} \frac{I_1 I_2}{R} \cdot L \quad \text{or, dividing by L}$$

$$f = \frac{\mu_0}{2\pi} \frac{I_1 I_2}{R} \quad \text{where } f = F/L \quad (5.1)$$

and $\frac{\mu_0}{2\pi} = 2 \times 10^{-7}$. Again it is possible to break Equation 5.1 into two parts: first the current I_1 creates a magnetic field B around it:

$$B = \frac{\mu_0}{2\pi}\frac{I_1}{R} \quad \text{magnetic field due to wire} \quad (5.2)$$

and then the force of the intermediary field B on I_2 is

$$f = B * I_2 \quad \text{force on a current in a B-field} \quad (5.3)$$

In equations 5.2 and 5.3, B is the magnetic field, measured in tesla (T).

There are several obvious differences between the above magnetic field formulas and the electric field formulas of Chapter 4:

1. We have replaced charges with currents.

2. The constants k and $\mu_0/2\pi$ are different for electric and magnetic forces

3. The distance dependence follows $1/R$ instead of $1/R^2$.

4. The force is computed per unit length of wire carrying the current, i.e., in terms of f = F/L.

There is another big difference, not yet indicated in the above list, which is the remarkably different vector nature of the magnetic field. One consequence of this is, unlike the electrostatic force, the force between parallel wires carrying currents in the *same* direction is *attractive*(!).

Before getting into the details of the B-field and its vector complexities, take a look at Figures 5.1 and 5.2 which make a pictorial comparison between electric and magnetic fields, forces, etc. The overview illustrated in the figures is that there is an analogy between the electric and magnetic force equations, force-fields, and the important example of a device that provides a uniform field. The device was a parallel plate capacitor for a uniform electric field; and a solenoid for a uniform magnetic field. We develop the electric and magnetic principles independently at first. Later, analysis of the interaction between them explains (1) the motor/generator, and (2) electromagnetic waves such as light.

The Magnetic Field

Coulomb's Law

$$\vec{F} = k \frac{q_1 * q_2}{R^2}$$

$$\vec{F} = q_2 * \vec{E}$$

E-Field

$$\vec{E} = k \frac{q_1}{R^2}$$

$$E = -\frac{\Delta V}{d}$$

Gauss' Law

$$AE = 4\pi k \sum q$$

$$C = \frac{A}{4\pi k d}$$

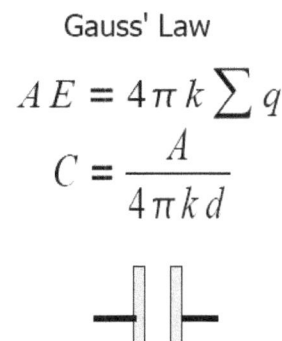

FIGURE 5.1 Illustration of electric force, field, and capacitor which has uniform field

Magnetic Force

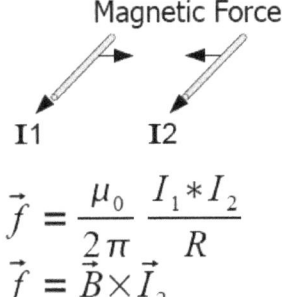

$$\vec{f} = \frac{\mu_0}{2\pi} \frac{I_1 * I_2}{R}$$

$$\vec{f} = \vec{B} \times \vec{I}_2$$

B-Field

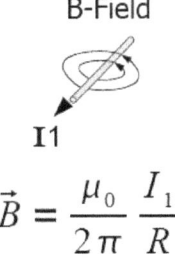

$$\vec{B} = \frac{\mu_0}{2\pi} \frac{I_1}{R}$$

Ampere's Law

$$2\pi r B = \mu_0 \sum I$$

$$L = \mu_0 \frac{N^2 A}{l}$$

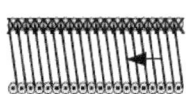

FIGURE 5.2 Magnetic force, field, and solenoid which has uniform B-field

5.1 Force between parallel Currents

As mentioned above, the constant in the Force equation 5.1 is $\frac{\mu_0}{2\pi} = 2 \times 10^{-7}$. The currents I_1 and I_2 are measured in amperes, the distance R is in meters, and f will have the units Newtons/meter. The force is attractive if the currents are parallel, and repulsive if the currents are antiparallel. The following examples will show that ordinary current flow such as that in power lines, etc., results in rather small forces.

Examples of Forces between Currents

1. Question: Two power lines carry current in opposite directions. The current in each power line is 500 A, the separation between the lines is 0.3 m, and the length of the lines is 60 m. What is the force between them?

    Answer: The force is $f * L = (2 \times 10^{-7} \cdot 500 \cdot 500/0.3) * 60 = 10$ N. The force pushes the two lines slightly apart. The force is equal and opposite on the two power lines.

2. Question: A lightning bolt splits into two parallel paths carrying currents of 500 A and 800 A in the same direction. The separation between the two paths is 6 cm, and the length of parallel paths is 10 m. What is the force between the two lightning paths?

    Answer: The force is $f * L = (2 \times 10^{-7} \cdot 500 \cdot 800/0.06) * 10 = 13.3$ N. The force is attractive, tending to push the two alternate paths back together.

3. Question: Three wires - P, Q, and R - form the 'L'-pattern shown in the Figure 5.3. Assume all three currents equal 20 A. What is the force per unit length on the wire at vertex Q of the 'L'?

    Answer: Like the case of several point charges, multiple wires exert forces on each other in pairs. Wire Q exerts a force on wire P. By Newton's Third Law, wire P exerts an

The Magnetic Field 73

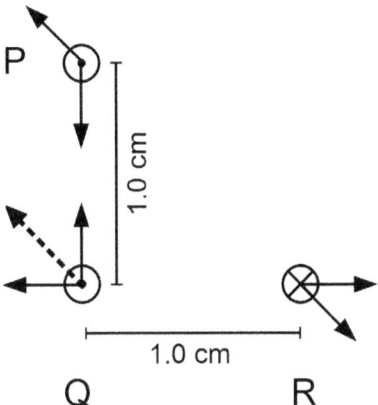

FIGURE 5.3
Magnetic forces between three wires. The dotted arrow shows the resultant force on wire Q.

equal and opposite force on wire Q. Similarly, R exerts a force on Q and vice versa. The two forces acting on Q add as vectors. The force diagram in the figure indicates how to add the two forces on Q. The result is the vector sum of equal leftward and upward forces, each having a value $\frac{\mu_0}{2\pi} \times 20 \times 20/.01 = .008$ N/meter. The net force is .011 N/meter, pointing 'Northwest'.

5.1.1 B-field Due to a Wire: Direction

The magnitude of the B-field due to a current-carrying wire was given above in Equation 5.2. The question is, what is the direction of B? If the direction of the B-field due to I_1 were radial, like the E-field, it would be impossible to make the force on test current I_2 depend on the direction of I_1. That is, if I_1 produces a field that is radially outward from the wire, how can this change when the current I_1 is turned upside-down? See Figure 5.4. The way out of this dilemma is to create a B-field that circulates around the current, as shown in Figure 5.5. The direction of this circulation clearly can change when I_1 flips direction. It introduces the complication, however, that the force on I_2 must be perpendicular to the B-field!

FIGURE 5.4
The absurdity of a radial B-field: a) currents attract; b) currents repel. How can the force on I_2 change when neither the B-field nor I_2 have changed?

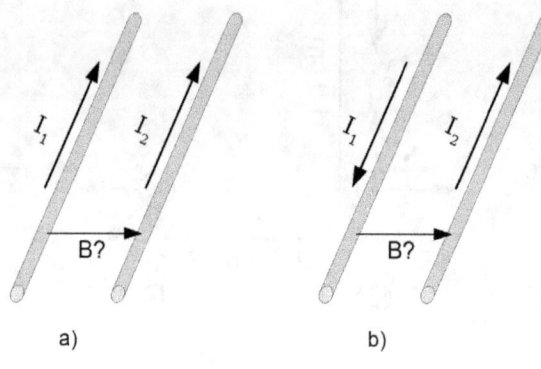

FIGURE 5.5
a) The B-field circulates clockwise around the I_1. B points downward where it hits I_2. b) The B-field circulates CCW because I_1 has flipped, taking B with it. Now B points *upward* where it hits I_2. Therefore the force F on I_2 has reversed.

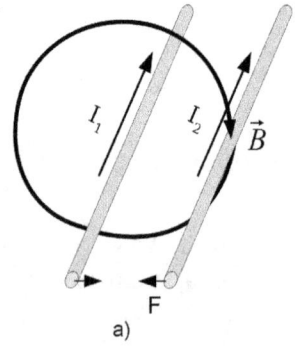 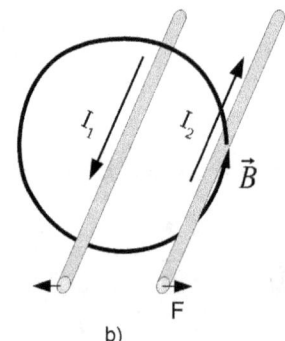

5.1.2 Righthand Rules

There are two (or so) righthand rules for determining the vector direction of magnetic fields and their effects. We need two: one for the direction of the field circulating around a current; and a second for the direction of the force due to the B-field acting on a second current.

Righthand Rule for a Wire

If a person grabs a wire carrying a current with the right hand, with the thumb pointing in the direction of the current, then the fingers wrap around the wire in the direction that the B-field circulates. Try this out on the left wire in Figure 5.5(a) and (b), and prove to yourself that the direction of the magnetic field due to I_1 conforms to this rule.

B-Field Examples

1. Question: A long, straight wire carries 40 A. How large is the magnetic field 8 cm from the wire?

   Answer: $B = (\mu_0/2\pi)^* I/R = 2 \times 10^{-7} * 40/.08 = 10^{-4}$ T.

2. Question: A long, straight wire carries a current towards the North. What is the direction of the B-field directly above the wire?

   Answer: Looking toward the north, the B-field circulates clockwise. Directly above the wire, the B vector points eastward.

3. Question: Two parallel wires are 10 cm apart and carry current of 20 A, as shown in figure 5.6. What is the B-field at the point A, halfway between the two wires?

   Answer: The B-field at point A is the vector sum of the B-fields from the two wires. The value of the B due to wire L is $B = (\mu_0/2\pi)^* I/R = 2 \times 10^{-7} * 20/.05 = 8.0 \times 10^{-5}$ T. The B-field due to wire L points up at point A. The field

FIGURE 5.6
Dot in circle = wire carrying positive current out of the page. Cross in circle: current into the page.

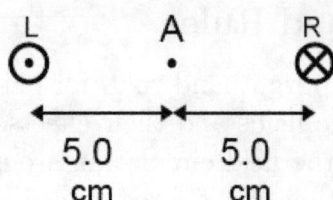

from wire R has the same magnitude, and also points up at point A. The vector sum is therefore B = 1.6×10^{-4} T.

Righthand Rule for Force on a Current

If a person points the fingers of the right hand in the direction of the current **I**, and then twists the hand so that the **B**-field points out of the palm, then the thumb will point in the direction of the force **F**. A mnemonic for remembering this sequence is '**I B**elieve in the **F**orce'! You can be sure the palm is facing correctly because the fingers of the right hand can bend at the knuckles from the current direction into the B-field direction by closing the hand. The Righthand Rule for the Force on a Current is illustrated in Figure 5.7.

FIGURE 5.7
Righthand Rule for the magnetic force on a current. Note: the circled-cross next to **B** symbolizes the B-field is a vector pointing into the page.

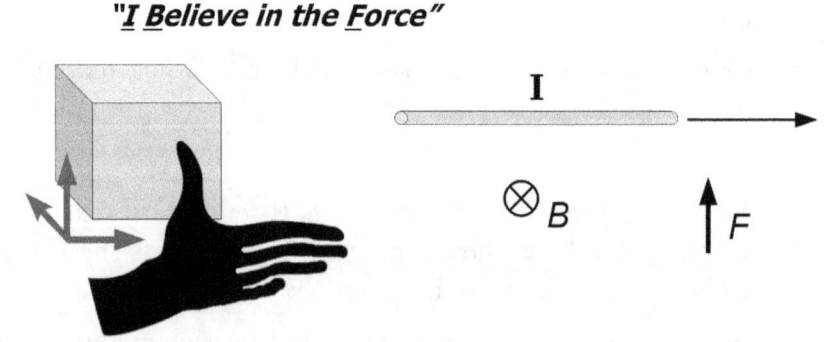

Magnetic Force on a Current Examples

The Magnetic Field

1. Question: A powerful permanent magnet produces a field of 0.5 T. A wire carries a current of 20 A at right angles to the field. The length of wire immersed in the field is 2 cm. What is the force on the wire?

   Answer: $f = F/.02(m) = B*I$, so $F = .02*.5*20 = .2$ N.

2. Question: The Earth's magnetic field is about 5.5×10^{-5} T. If wire carries a 300 A current perpendicular to the Earth's field, what is the magnitude of the force per meter of wire?

   Answer: The Earth's magnetic field points north. The force per meter is given by $f = B*I = 1.65 \times 10^{-2}$ N/m.

5.2 Force on a Charge Moving in a B-field

The magnetic force on a current-carrying wire, Equation 5.3, gives the magnitude of the magnetic force on a wire due to charge moving inside the wire. The force on a wire is a specific case of the more general principle: When a charged particle moves in a path perpendicular to a magnetic field, there is a force at right angles to both the magnetic field and the path of the moving charge. What is the force on a charge q moving at speed v?

$$F = qvB \qquad \text{force on a charged particle in a B-field} \qquad (5.4)$$

5.2.1 Moving Charge Similar to Current

In this section we show that Equation 5.4 for a moving charge is equivalent to Equation 5.3. The following example illustrates the point:

Moving Charge Example
Situation:
   Suppose that during 1 second in a conducting wire, the current flow is made up of 10^9 electrons going a distance 1 cm. The B-field is perpendicular to the current flow, and has magnitude B Tesla (B not specified).

1. First according to 5.4 for charged particles: the total force is $F = 10^9 * evB = 10^9 * 1.6 \times 10^{-19}$ C $* 10^{-2}$m/s $* B = (1.6 \times 10^{-12} \times B)$ N.

2. On the other hand, treating this situation as the force on a wire, we need to know the current: 10^9 electrons moving in 1 second is defined as a current of $I = 10^9 * 1.6 \times 10^{-19}$ C$/(1s) = 1.6 \times 10^{-10}$ A. Then, according to the wire equation 5.3, $F = f * 10^{-2}$ m $= B * 1.6 \times 10^{-10}$ A $* 10^{-2}$ m $= (B \times 1.6 \times 10^{-12})$ N.

So the two points of view for magnetic force give the exact same result.

5.3 Current versus Moving Charge in B-field

A wire experiences a force in a magnetic field because charge inside the wire is moving across the B-field. However, the path of a free particle is different from charge in the wire because the wire forces it to move along the length of the wire. A free particle is accelerated in the direction of the magnetic force — i.e., perpendicular to the particle's velocity. This leads the particle to follow a circular path. Consider Figure 5.8. Assume the particle has

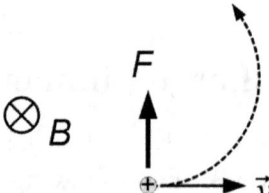

FIGURE 5.8 Charged particle moving in a B-field follows a circular trajectory.

mass m, and charge q. The force of equation 5.4 is $F = qvB$. This must equal the mass times the centripetal acceleration, so

$$qvB = \frac{mv^2}{R} \quad \text{and, solving for R}$$
$$R = \frac{mv}{qB} \tag{5.5}$$

The Magnetic Field

Here R is the radius of the circular trajectory followed by the particle. The basic result of equation 5.5 is that the circle gets larger for larger mass and velocity, and smaller for increasing charge and B-field. The radius is a way to identify subatomic, atomic and molecular particles. A typical application is the *mass spectrometer*. Usually the charge on a particle to be measured is one or a couple of electronic charges. The spectrometer has a known magnetic field. The particle enters the spectrometer with a known velocity. The spectrometer is set up to measure the radius, and this determines the mass of the particle. The mass is an excellent guide to the identity of a particle. For complex organic molecules, where many possible structures can have the same mass, the different chemicals in a sample may be separated first by gas chromatography (GC), then broken into smaller pieces and submitted to the mass spectrometer to identify all the pieces. This tool is called a *GC Mass Spec*, or *GCMS* and is very important for identifying biological compounds, in forensic investigations, and environmental studies. In the case of elementary subatomic particles, a *cloud chamber* or *bubble chamber* type of mass spectrometer is used to visualize the trajectory and then determine the charge-to-mass ratio. A common type of mass spectrometer is used for vacuum and pipeline leak detection, because any openings can be located by sensitive detection of a tracer gas that is recognizable from its molecular mass.

Circular Trajectory Examples

1. Question: A sample containing C^{12} and C^{14} enters a mass spectrometer. The ions are each charged with a single electronic charge, and they both have velocity 2×10^5 m/s. The mass spec uses a magnet with a B-field equal to 0.2 T.

    (a) What is the radius of the path of the C^{12} atom?
    Answer: C^{12} has a mass of 12 amu, where 1 amu = 1.66×10^{-27} kg. Therefore the radius is $R = mv/(qB) = 12 \times 1.66 \times 10^{-27} * 2 \times 10^5 / (1.6 \times 10^{-19} * 0.2) = 0.125$ m.

(b) What is the radius of the C^{14} trajectory?

Answer: The C^{14} has a mass of 14.00 amu (atomic mass units), while the C^{12} has a mass of 12.00 amu. Therefore the radius of the C^{14} will be $(14/12) \times 0.125$ m = 0.146 m.

(c) What is the energy change of the C^{12} ion due to the force of the B-field?

Answer: There is no change in energy of a particle moving in a circular trajectory because the magnetic force is at right angles to the motion of the particle.

5.3.1 Angle Between B-field and Current Affects Force

Equations 5.3 and 5.4 predicted the force of a magnetic field on a current or moving charged particle when the B-field is perpendicular to the current or velocity of the charge. If the B-field is *not* perpendicular to the motion of the charge, then there is an additional factor $\sin\theta$ that multiplies the force. Equations 5.3 and 5.4 become:

$$f = BI * \sin\theta \quad \text{force on current, B at angle } \theta \text{ to I} \quad (5.6)$$

and

$$F = qvB * \sin\theta \quad \text{force on charge, B at angle } \theta \text{ to } \vec{v} \quad (5.7)$$

Some possible configurations appear in Figure 5.9a-d. In Figure 5.9a, $\theta = 90°$, $\sin\theta = \sin 90° = 1$, and so there is no change from Equations 5.3 and 5.4. Figure 5.9b is another important case of B-field direction parallel to the direction of the current. This configuration gives zero magnetic force on the current, because $\sin\theta = \sin 0° = 0$. In the case of a charged particle, the particle will continue in a straight line at constant velocity, parallel to the field.

In Figures 5.9c and d, there is an acute angle θ between B and the current. Then the magnitude of force F is $BI\sin\theta$. The

The Magnetic Field

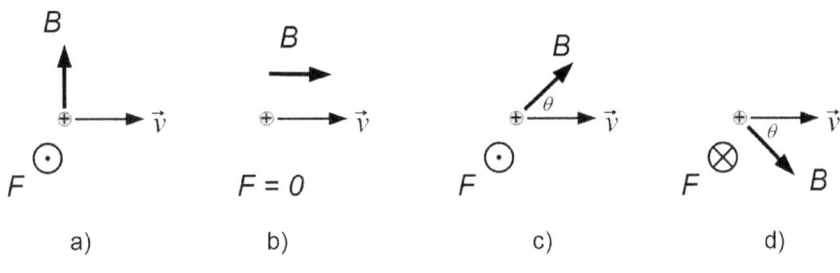

FIGURE 5.9 Magnetic force F on a moving charge, depending on B-field direction: B-field is (a) perpendicular to, (b) parallel to, (c) and (d) at angle θ to, velocity.

direction of the force, into the page or out of the page, depends on whether the B vector is clockwise or counter-clockwise, resp., relative to the particle direction. The right hand rule correctly defines the direction of the force: the thumb direction reverses in (d). This is necessary to allow the fingers to close slightly as they bend at the knuckles through the acute angle θ, and go from pointing in the direction of the current to pointing in the direction of the B-field.

Example of B-field and Current at an Angle

- Question: In the northern hemisphere, the magnetic field actually points downward at an angle of *declination* that depends upon our exact location. Suppose in Phoenix, a wire carries current of 300 A northward. Then the angle between the Earth's field and the wire is the angle of declination in Phoenix (59 degrees). What is the force on each meter of wire?

  Answer: According to Equation 5.6, the force per meter is $f = BI\sin\theta = 5.5 \times 10^{-5} * 300 * \sin 59 = .014$ N/m. The right hand rule gives the direction of the force as west, which is perpendicular to both the direction of the I (north) and the direction of B (in between north and downward). With the fingers pointing north, and the palm down, the thumb points west. The fingers can then bend from north, down through the angle of declination, to the direction of B.

5.4 Magnetic Torque on a Current Loop

Consider the B-field force on a rectangular loop that carries a circulating current. See Figure 5.10. This is a very common configuration in many kinds of electric motors. According to Equation 5.6, the forces on the current in opposite sides of the rectangle are opposite in direction because the current changes sign. In other words, $\sin\theta$ reverses sign for the opposite flows of current. The opposing forces on two sides of a rectangular loop constitute a torque on the loop. This section shows how a B-field that is spatially uniform can place a torque on the current it induces. Figure 5.10 shows a current loop standing upright in a uniform magnetic field, with I amperes flowing through the loop. According to Equation 5.6, the force due of the B-field acting on vertical

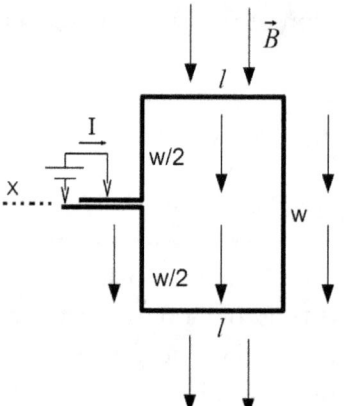

FIGURE 5.10
A rectangular current loop immersed in a uniform B-field.

segments of wire in the loop is zero, because B and I are aligned so $\sin\theta = 0$. The top and bottom segments experience a force of magnitude $F = BIl$, since B is perpendicular to the wire. Here $\sin\theta = 1$. The right hand rule says that the force on the top segment of the loop is back into the page. The force on the bottom segment is out of the page. This pair of opposed forces each has a lever arm $w/2$. Added together, they give a torque $\tau = IwlB$

about the axis x. This is the maximum torque experienced by the loop— the B-field lies in the plane of the loop, as in Figure 5.10.

Imagine the loop rotates under the influence of the above torque. If the plane of the loop is now tilted back by an angle Θ, as in Figure 5.11, the forces on the horizontal wire segments

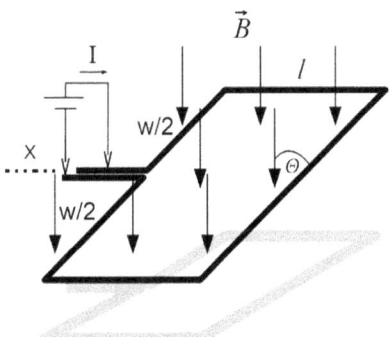

FIGURE 5.11
A rectangular current loop tilted back an angle Θ.

are the same but the lever arm decreases due to a factor of cos Θ, where Θ is the angle between the B-field and the plane of the loop. Note that when tilted, the forces on the left and right rectangular segments in Figure 5.11 now become no longer zero, but oppose and 'line up', and so they still produce neither a net force nor a net torque on the loop.

Considering, generally, a coil with N turns:

$$\tau = NIAB \quad \text{maximum torque}$$
$$\tau = NIAB\cos\Theta \quad \text{torque vs. angle} \qquad (5.8)$$

5.5 Ampere's Law

We developed Gauss's Law in section 4.5 in order to handle a number of electrostatic configurations, especially uniform electric fields. The same basic technique can apply to uniform magnetic fields. As in the electric case, where the capacitor provides a uniform electric field, there is a magnetic device, the *solenoid*, which produces a strong, uniform magnetic field. A solenoid is just a long coil of wire. Ampere's Law provides a method for finding the B-field in a solenoid, based on the geometrical configuration and the current going through the solenoid. Just as Gauss's Law derives from E-field emanating from a point charge, the key to deriving Ampere's Law is to look at the magnetic field that circulates around a wire.

We have seen from equations 5.2 that the B-field around a wire is proportional to current I and inversely proportional to radius R. The B-field circulates around a long, straight wire. This is a common geometry. If the wire is wrapped into a long helical coil, or solenoid, this configuration of current produces a uniform field. We want to *prove* the field inside a solenoid is uniform, and then find how the strength of the B-field depends on the current and geometry of the coil.

The key to Ampere's Law is that the overall strength of circulation of the lines of force around a current remains the same when the geometric shape is changed from a straight wire to a solenoid. Ampere's Law is as follows:

- Consider a collection of wires surrounded by an imaginary, closed path. Then the strength of circulation of the B-lines of force surrounding the wires is equal to

$$K = \mu_0 I_{total} \qquad \text{Ampere's Law} \qquad (5.9)$$

where I_{total} is the total net current piercing the area within the imaginary path.

What exactly do we mean by the *circulation*, K, of the B-field around the path? The concept is similar to the circulation of a fluid. If a fluid circulates, say in a whirlpool, then an object carried by the fluid can gain energy by following a complete circular path that encloses the vortex of the whirlpool. Then the current will always be pushing the object along the direction it is moving. Similarly, K accumulates wherever the B-field is aligned with the closed path chosen to evaluate K. If the B-field is not parallel to a segment of the path, then only the component aligned with the path contributes to K. If the B-field is perpendicular to a segment of the path, then there is no contribution to K along that segment. If the closed path encircles a wire carrying current, then there will be a non-zero accumulation of K along most or all of the path.

5.5.1 'Proving' Ampere's Law for a Straight Wire

Consider a wire carrying current I, and an arbitrary closed path around the wire. The circulation is the product of the length of the path around the wire times the strength of the magnetic field aligned with the path. In particular, choose a path which is a circle of radius R that is centered on the wire. The wire has current I. Then everywhere along this path, the value of B is $\mu_0 I/2\pi R$. The B-field is perfectly aligned with the path. Therefore K is just the circumference of the path, $2\pi R$, multiplied by the B-field value, or $K = (2\pi R) \times (\mu_0 I/2\pi R) = \mu_0 I$, proving Ampere's Law, equation 5.9.

5.5.2 Solenoid B-field

The field due to a straight, current-carrying wire consists of concentric B-field circles centered on the wire. If the wire is now bent into an arc, the B-field becomes weaker on the outside of the curve, and stronger on the inside of the curve, as shown in Figure 5.12. If the wire is bent completely into a loop, or better, into a helical coil, then the field becomes very strong inside the coil, and very weak outside the loop. Now to determine the field inside a long solenoid. Ampere's Law will be applied. In Figure 5.13 only the central portion of a much longer solenoid is shown. There is no net circulation contributed along the top t and bottom b paths, because the B-field must be vertical in the picture, by symmetry. Contribution to circulation along the left path, o, is vanishingly small because this path is far outside the solenoid. Furthermore, B-field contributions along the path, o, tend to cancel out because the wire segments on the left and right half of the coil, pointed out of and into the page, resp., make opposing contributions to

FIGURE 5.12 Lines of force around a wire (a), distort when the wire is bent (b). In the case of a helix, the wire is bent again and again, concentrating the B-field inside the coil.

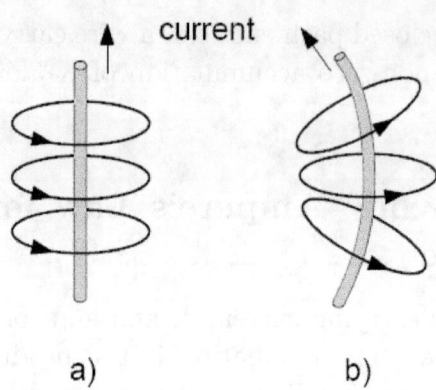

the field outside of the solenoid. Ampere's Law calculation of

FIGURE 5.13 Path for Ampere's Law is used to evaluate B-field inside the solenoid.

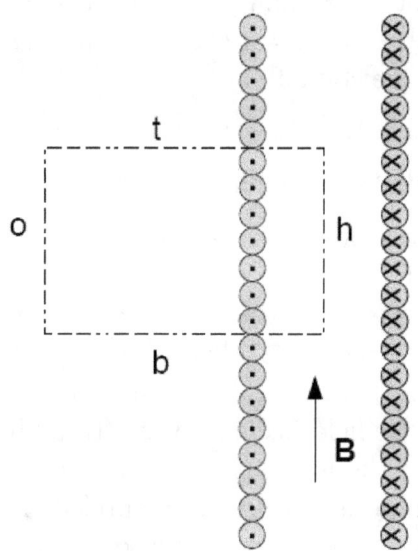

Equation 5.9 is therefore $BL = \mu_0 I N_{tpm} L$, where L is the height of the imaginary path segment h, and N_{tpm} is the number of turns per meter wound around the coil. Therefore B inside the solenoid is finally,

$$B = \mu_0 I N_{tpm} \qquad (5.10)$$

5.6 Magnetic Materials

In the discussion so far, the magnetic fields and forces depended on currents in wires or currents due to free charges. These give rise to magnetic fields, and in turn, currents and moving charges experience a force when immersed in a magnetic field. It is very important also to consider microscopic currents in materials. Materials contain electrons, which are inherently charged. Because electrons orbit atoms and have an intrinsic spin, the charge in electrons is in continually circulating motion. These microscopic currents are little loops giving rise to microscopic magnetic fields, and to atomic-scale torques on the electrons when external magnetic fields are imposed.

We have seen that electric fields also influence the individual charges in materials. The electric field can displace positive charges relative to negative charges. In metals, this causes significant macroscopic electric currents to flow in circuits. The molecules of a nonmetal also become somewhat polarized in the presence of an electric field. Or, as in the case of H_2O which has its own dipole moment, the electric field can orient the molecules in the direction of the field. But in neither metallic circuits nor in electrically polarized materials is it ever possible to accumulate a large concentration of positive or negative charge. Electrostatic forces generally remain small and limited to a laboratory curiosity. Charge accumulation only arises from the actual transport of charged objects, as in the case of lightning or the van der Graaf generator.

Magnetic effects on matter can be much more dramatic than electric polarization effects. What is special about magnetic polarization is that in a few materials, known as *ferromagnetic* materials, the atomic currents strongly influence neighboring currents to circulate all in the same direction. A thorough explanation of ferromagnetism goes way beyond the scope of undergraduate physics, and has been an area of theoretical inquiry for more than a century. The important point is that ferromagnetism is a strongly cooperative effect among the electron spins. The atomic current-loops align over large regions called *domains*. The intrinsic magnetic field which the electrons generate in these domains is big, and a relatively smaller external magnetic field is enough to turn and align the direction of the intrinsic B-field in a domain. The field produced by the domain is parallel to, and much larger than, the applied field! Ferromagnetic materials therefore amplify the fields produced by the currents in wires and coils wrapped around such materials. Furthermore, the force produced by an external field on a current-carrying coil is strongly enhanced if the coil is wrapped around a ferromagnetic core.

In summary, ferromagnets can create large B-fields, which can exert powerful forces and torques on current-carrying coils. This is the basis for powerful electric motors, which have been one of the greatest technological factors in industrial progress in the 20th century. We will see, in the chapter on Faraday's Law, that the same ingredients are used to make efficient generators for converting mechanical energy into electricity.

The most common ferromagnet is iron. Practical ferromagnetic materials include alloys of iron, cobalt, and nickel. Permanent magnets are possible if domains are aligned in a magnetic field, and then locked in place by crystal defects. Extremely powerful permanent magnets are made from combinations of the above elements with rare earth elements such as Neodymium.

Chapter 6

Faraday's Law

Objective: The magnetic fields discussed so far have been static, meaning they do not change in magnitude or direction. Next we explore the effect of a changing magnetic field, using Faraday's Law and Lenz's Law. A changing magnetic field causes an electromotive force, or emf. Faraday's Law governs the size of the emf, and Lenz's Law provides the direction. An important application of these principles is that we can turn an electric motor into a generator of electrical power.

Chapters 4 and 5 showed that Electric and Magnetic fields can exert forces on charges and currents, respectively. A practical example is the electric motor, where a strong magnetic field applied to a current-carrying coil produces a powerful torque in an electric motor. Such motors convert electric energy into mechanical energy. In this chapter, we find that the transfer of power from electrical sources to mechanical applications is reversible: We can use the same coil and magnetic field. The mechanical power can either move the coil within the magnetic field, or move the magnetic field within the coil. In either case the coil becomes energized with an *emf*, which can drive a current through a load. Thus the coil moving within the magnetic field converts mechanical power back into electrical power. Every motor is also a generator.

6.1 EMF due to Changing B-Flux

Figure 6.1 depicts a circular conducting loop immersed in a changing B-field. The B-field is oriented at right angles to the plane of the loop. Snapshots (a) and (b) show the B-field at two different times. The coil and field are fixed in position but the B-field decreases between time (a) and time (b). The *B-flux* is the B-field multiplied by the area of the loop, and has the units, Webers (Wb). A flux of 1 Weber corresponds to a magnetic field of 1 Tesla penetrating an area of 1 m², or 1 T-m² =1 Wb. The flux is the number of B-lines of force penetrating the loop. The B-flux is likewise at right angles to, and pointed into, the plane of the page. Therefore, the B-flux also decreases between time (a) and time (b). According to Faraday's Law, the B-flux changing in the

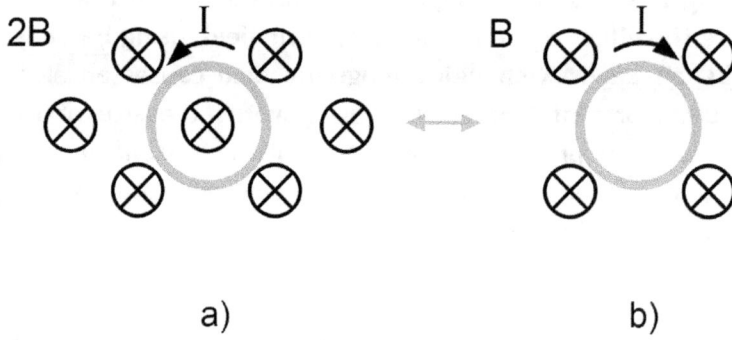

FIGURE 6.1 Circular conducting loop experiencing a decreasing field, (a) ⟶ (b), or an increasing field, (b) ⟶ (a).

loop generates an *emf* around the loop given by

$$emf = -\frac{\Delta \Phi}{\Delta t} \qquad \text{Faraday's Law} \qquad (6.1)$$

where $\Delta\Phi/\Delta t$ is the rate of change of B-flux. The negative sign gives an indication of the direction of the *emf*. The *emf* attempts to create a circulating current that will oppose or even cancel the change in flux applied by the external B-field. Therefore in Figure 6.1, as the B-field decreases from a) to b), this *induces* a clockwise current. By the right-hand rule, such a current tends to maintain the amount of B-flux through the coil in a direction into the paper. Similarly, as the B-field increases from b) to a), this induces a counter-clockwise current. The negative sign embodies *Lenz's Law*.

> **Lenz's Law**: Any change in magnetic flux through a conducting loop causes an *emf* in the loop which would tend to oppose that change in magnetic flux.

Faraday's Law Example

- Question: A 20-turn coil has an area 10 cm². If the magnetic field increases from 0 to 0.5 Tesla through the coil in 1/60 second, what is the *emf* generated in the coil?

  Answer: Each turn of the coil builds an *emf* given by Faraday's Law. These voltages all add in series. Therefore Equation 6.1 becomes
  $$emf = -N\frac{\Delta \Phi}{\Delta t}$$
  where N = 20 is the number of turns in the coil. The voltage is then $emf = 20 \times 0.5 \text{ T} \times 60 \text{ s}^{-1} \times 10^{-3} \text{m}^2 = 0.6$ V.

Figure 6.2 shows a circular conducting loop moving toward a fixed B-field. When the loop crosses into the B-field, B-flux increases through the loop. This also induces an *emf* in the coil. The direction of the emf is such that it creates a current I which tends to 'fight' the increase in B-flux in the direction into the page.

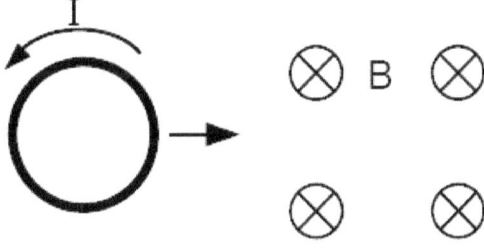

FIGURE 6.2 Circular conducting loop moving into a fixed B-field. The B-flux through the loop will increase, inducing a current I which will oppose the increase.

Figure 6.3 shows a region of B-flux moving toward a stationary wire loop. When the B-field crosses into the loop, B-flux increases through the loop. This again induces an *emf* in the coil. The direction of the emf is such that it creates a current I which again tends to 'fight' the increase in B-flux going into the page.

FIGURE 6.3 B-field moving into a circular conducting loop. The B-flux through the loop will increase, inducing a current I which will oppose the increase.

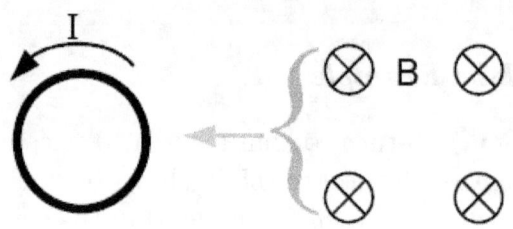

Figure 6.4 shows a fixed B-field and a wire loop, immersed in the B-field, rotating around a vertical axis. When the coil rotates from orientation b) to orientation a), B-flux through the loop decreases, because the area of the coil presented to the B-field decreases. This again induces an *emf* in the coil. The direction of the *emf* is such as to create a current I tending to 'fight' the decrease in B-flux going through the loop.

FIGURE 6.4 Circular conducting loop rotating in a fixed B-field. As the loop rotates, the B-flux going through it into the page decreases, and induced current I opposes the decrease (left).

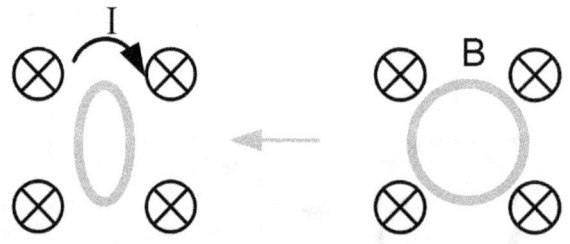

Faraday's Law

Figure 6.5 shows a wire hairpin, with a cross bar completing the loop, immersed in a fixed B-field. When the crossbar moves to the right with velocity v, the flux in the loop increases because the area of the loop increases. This again induces an *emf* in the coil. The direction of the *emf* is such as to create a current I tending to 'fight' the increase in B-flux going through the loop. The rate of increase of area is $\Delta A/\Delta t = v \times h$, so the $emf = -\Delta \Phi/\Delta t = Bvh$.

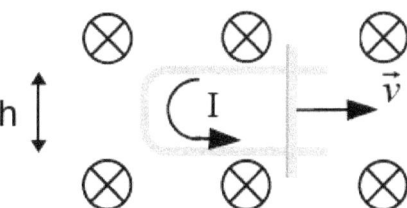

FIGURE 6.5
Hairpin loop with increasing area in a fixed B-field. As the crossbar moves right, the flux through the loop increases, inducing current I.

More Examples of Faraday's Law Induction

1. Question: In a solenoid having 10,000 turns and 10 cm² cross-sectional area, a B-field of 0.05 T is aligned along the axis. The field is held constant in size and fixed in direction. In 0.3 second the solenoid flips around so its long axis points in the exact opposite direction. What is the size of the *emf* generated in the solenoid?

    Answer: When a coil flips 180°, the flux through it does not change in magnitude, but it does change in direction. The change in flux is twice its magnitude. Therefore $\Delta \Phi = 2 \times 0.05$ T$\times 1.0 \times 10^{-3}$ m² $= 1.0 \times 10^{-4}$ Wb, and $emf = -N\Delta\Phi/\Delta t = 10^4 \times 1.0 \times 10^{-4}$ Wb $/ 0.3$ s $= -3.33$ V. The $-$ sign in the answer indicates the direction, and has no meaning here because the problem does not specify the geometry.

2. **Question:** A wire bent into a large, circular loop slides across the face of a magnet. At the beginning of the motion, the magnetic field in the loop is zero, and at the end of the motion the loop circles the entire pole face of the magnet. The loop area is 300 cm². The area of the pole faces is only 100 cm². The B-field of the magnet is 1.2 T, assumed constant across the pole face of the magnet. If the loop takes 0.5 s to complete this motion, what is the average *emf* induced in the loop?

    Answer: The flux change in this problem is $B \times Area\ of\ Magnet = 1.2\ T\ \times 0.01\ m^2 = .012$ Wb. Therefore $emf = -\Delta\Phi/\Delta t = -0.012$ Wb / 0.5 s $= -0.024$ V. Again the $-$ sign is meaningless in the context of this problem.

6.2 Limit On the EMF – the Self-Inductance

It might seem from Faraday's Law Equation 6.1 that there is essentially no limit to power generation using changing magnetic fields, due to the following erroneous logic:

1. Apply a change in B-flux, resulting in the *emf* of Equation 6.1.

2. To maximize power, use a very low resistance coil, implying a very large current, I.

3. The power output will be the product I·*emf*, which can be as large as we please by making R very small.

The flaw in this logic is in step 1. It is possible to *apply* a B-field, but the actual B-flux experienced by the coil is the sum of the applied B-field *plus* any B-field the coil generates to oppose the attempted change. If R is large, the opposing B-field is small because the induced current is small, so the electrical power is small. As R is made smaller— and it can be made literally zero in a superconductor— the opposing current due to Lenz's Law grows large enough to cancel the change in flux completely. Then Equation 6.1 gives a vanishingly small *emf*— just so $I_{coil} = emf/R$ produces its own B-field to virtually cancel the applied B-flux. Now the current I_{coil} is large, the flux change in the coil is small, the *emf* is small, and the generated power is still small.

The ability of a superconductor to exclude a magnetic field is known as the *Meissner Effect*. There is a dramatic demonstration in which a small permanent magnet levitates above a cup made of superconducting foil. The demo is activated by lowering the superconducting material below its superconducting transition temperature, at which point the superconductor excludes the magnetic field and can actually raise the magnet until it floats above the superconductor.

6.3 Inductance

In the previous Section 6.2 the current induced in a coil reduced magnetic flux changes due to an applied B-field from an external source. This section pursues a related idea. When an external voltage change is applied suddenly to a coil, the current response is limited: the voltage does increase the current, but the B-flux change through the coil induces an *emf* that reduces the voltage change in the circuit and thereby slows down the current change through the coil. This is due to the *self-inductance*, or just plain 'inductance', of the coil. Inductance in a circuit prevents instantaneous current changes in response to rapid voltage changes.

6.4 Forces on Induced Currents – Magnetic Damping

When a B-field suddenly turns on near a loop, coil, or solid chunk of metal, the change in magnetic flux induces a current which tends to oppose the B-field. Previous sections discussed the induced *emf*. This section shows that the *emf* causes induced currents in the loop or solid conductor. These induced currents interact with the B-field, producing forces and torques on the loop or solid conductor. The forces and torques can be put to practical use to turn motors and to produce *magnetic damping*, a term which will be explained below.

For example, suppose the North pole of a magnet rapidly approaches a solid piece of highly conducting aluminum. The B-flux goes from zero to a large value in a short time. This induces currents, called 'eddy' currents, in the aluminum. The eddy currents try to counteract the change in applied field. The situation is depicted in Figure 6.6. The eddy currents produce a B-field di-

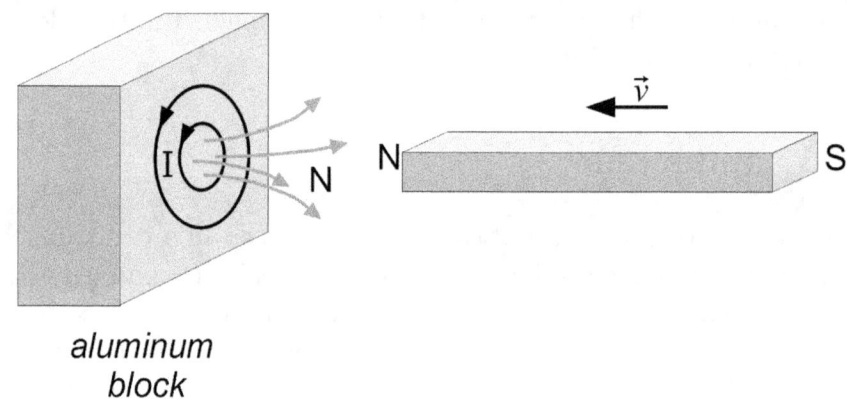

FIGURE 6.6
Eddy currents induced in a chunk of aluminum repel an approaching magnet. Only the surface currents are shown, but in fact eddy currents circulate a considerable depth into the metal.

rected out of the aluminum toward the North pole of the magnet. Since a diverging B-field should be associated with a magnetic North pole, it is as if the eddy currents are themselves a temporary solenoid, whose North pole is directly facing the North pole of the approaching magnet. Hence the two North poles repel. The approaching magnet is exerting a force on the block of aluminum. The block of aluminum exerts an equal and opposite force on the magnet.

On the other hand, suppose the magnet reverses direction, and the North pole of the magnet now pulls away rapidly. The eddy currents reverse, trying to restore the decreasing field from the receding magnet. The eddy currents now create a South pole, which attracts the magnet and tends to slow down its departure. We see that whatever motion the magnet has, the induced eddy currents oppose the motion. Assuming the magnet slows down, the rate of change of the field will decrease. According to Faraday's Law, the *emf*'s and the induced eddy currents also decrease. The magnet decelerates less rapidly.

The effect of the magnet on a good conductor is very similar to viscous damping. Like an object moving through oil, the relative motion of a magnet and a good conductor slows down regardless of the direction of motion. The faster the motion, the stronger the damping force. This is called 'magnetic damping', and can be used to bring moving machinery to a graceful stop, or, for another example, to arrest the oscillation of a mechanical weighing scale near it's equilibrium point. You may have been to an amusement park and noticed a slot between the rails of the roller coaster. When the roller coaster comes into the loading area to let off and take on passengers, a conducting plate under the cars passes in between the poles of a strong magnet. This brings the roller coaster to a smooth stop, without any moving parts or friction pads that could wear out.

Chapter 7

AC Circuits

Objective: We have previously described steady current flow, e.g., batteries driving current through various series and parallel circuits and producing power. This was <u>direct</u> current. It is not the majority of the electrical energy we use. The purpose of this chapter is to extend your concept of electricity to the very important circuits that provide most of your lighting, electrical transmission, manufacturing, data processing, and home equipment. This is <u>alternating</u> current. Ohm's Law will still apply for resistors. However, capacitors and solenoids have an interesting response to alternating current, and you will find an analogous current-voltage relationship for these circuit elements. The analog of resistance for capacitors and coils is called <u>impedance</u>.

7.1 AC Current

Alternating current is abbreviated as 'AC'. The power you use in your home is AC. Electrons are pushed first in one direction, then the opposite direction, through the various appliances in your home. The actual distance traveled back and forth by any one electron is small, because the number of electrons available in the metal wiring is enormous. So no single electron gets very far. However, the amount of electromotive force pushing on the charges is quite large— hundreds of volts. This force pushes back and forth to produce the alternating current.

Consider a typical home circuit, such as a hair dryer. At any instant in time, the current is moving in the direction that the *emf* is pushing it. Therefore, at any instant, the negative charges in the hair dryer are moving from a more negative to a more positive voltage and gaining energy from the power company. After gaining a little energy they bump into the impurities in the heating coils of the dryer and lose their energy to the atoms they hit, causing atomic vibration and a rise in temperature. So at different points in time, the voltage and the current *change* direction, but they change direction *together*. Therefore the flow of power is always the same: from the power company, into the heating coils, and thence to your hair.

In the following section we will express the current, voltage, and power in mathematical form. This chapter will then show how the AC behavior changes when capacitors and solenoids replace resistors, and the useful functions such circuits can perform.

7.2 Current-Voltage Relationship in a Resistor

We have already described the I-V (current-voltage) relation for a resistor in Section 3.5 Figure 3.3, reproduced here as 7.1. Figure 7.1 is different, however. The I-V plot now extends from the first quadrant, for positive voltage and current through the resistor, into the third quadrant, where the voltage applied is negative, and the resulting current is also negative. The figure also indicates the response to an applied sinusoidal AC voltage. At $t = 0$, both the current, I_{OUT}, and voltage, V_{IN}, are zero. Then the voltage swings toward positive values, and the current also swings to positive values. Similarly, the voltage then turns negative, and the current follows. Several cycles of voltage swing are shown. It is clear that $V = I * R$ holds at every point in time. In particular,

AC Circuits

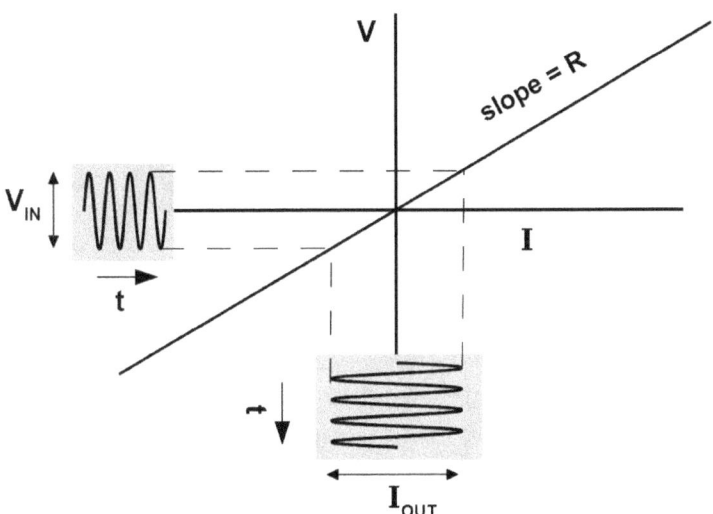

FIGURE 7.1
AC Ohm's Law I-V curve shown with voltage input, V_{IN}. For each voltage value there is a corresponding current value. The current time-dependence is I_{OUT}, shown in the vertically-oriented plot against time.

the amplitude values, I_{peak} and V_{peak} obey the same Ohm's Law equation.

7.2.1 Math Representation of AC

The voltage input to Figure 7.1 is a sine wave function of time. If the frequency of the sine wave is f, the voltage has the mathematical representation,

$$V_{IN} = V_{peak} \sin(2\pi f t) \qquad (7.1)$$

In other words, in the first second, the argument of the sine function runs from 0 up to $2\pi f$. Each 2π change in the argument causes the sine function to swing through another complete cycle. Therefore there are f complete cycles in the first second and each additional second.

60 Hz House Current Example

- Question: 'House current' refers to 120 VAC that comes from the wall outlets in your house. What is the mathematical expression for house current?

Answer (part I): The general mathematical expression for AC voltages is $V_{peak}\sin(2\pi ft)$. We have $f = 60$ Hz. Therefore the sine portion of the expression is $\sin(2\pi ft) = \sin(377t)$. In the next section we will define VAC, and this will allow us to evaluate V_{peak}.

7.2.2 AC Power in a Resistor

Instantaneous AC Power

The power that dissipates when an AC voltage is applied to a resistor follows the basic energy argument of Section 3.2 The direct current equation (3.2) applies at each instant of time. So for a time varying voltage, one could evaluate the *instantaneous* power as the product of the instantaneous voltage and instantaneous current. Hence the instantaneous power for sinusoidal AC voltage and current is

$$\begin{aligned} P &= VI \\ &= V_{peak}\sin(2\pi ft) \cdot I_{peak}\sin(2\pi ft) \\ &= V_{peak} \cdot I_{peak}\sin^2(2\pi ft) \end{aligned} \quad (7.2)$$

Average AC Power

Therefore the electrical power in a resistance, as given by Equation 7.2, oscillates rapidly between zero and $V_{peak} \cdot I_{peak}$ Watts, because $\sin^2(2\pi ft)$ oscillates between 0 and 1. However, the instantaneous power in a hair dryer, motor, and many other devices is much less interesting than the average power. You cannot detect the instantaneous heat pulses from a hair dryer because it takes thousands of cycles for the hair dryer to reach operating temperature. The reason is the specific heat of the heating element— the heat in a single cycle of AC power is insufficient to make much change in the temperature. So what is the average power represented by Equation 7.2?

Looking just at the function $\sin^2(2\pi ft)$, we can use the trigonometric identity, $2\sin^2\theta - 1 = \cos 2\theta$. Now take the average of both

AC Circuits

sides of this equation over any complete number of cycles. Here $\langle x \rangle$ indicates the average of x.

$$\langle 2\sin^2\theta \rangle - 1 = \langle \cos 2\theta \rangle = 0 \quad \text{and re-arranging,}$$

$$\langle \sin^2\theta \rangle = \frac{1}{2} \tag{7.3}$$

In the above, we took $\langle \cos 2\theta \rangle = 0$ because the cosine function has equal positive and negative excursions, so averages to zero.

Now combining Equations 7.2 and 7.3, the average power is

$$\begin{aligned}
\langle P \rangle &= \frac{1}{2} V_{peak} \cdot I_{peak} \quad \text{or using } V = IR, \\
&= \frac{1}{2} \frac{V_{peak}^2}{R} \quad \text{and} \\
&= \frac{1}{2} I_{peak}^2 \cdot R \\
&= \frac{1}{2} P_{peak}
\end{aligned} \tag{7.4}$$

Equations 7.4 show that in an AC circuit with a resistive load, the average power is 1/2 the peak power.

We can make Equations 7.4 more familiar if we define the AC voltage and current as follows:

$$\begin{aligned}
V_{rms} &= V_{peak}/\sqrt{2} \quad \text{definition of } rms \text{ voltage} \\
I_{rms} &= I_{peak}/\sqrt{2} \quad \text{definition of } rms \text{ current}
\end{aligned} \tag{7.5}$$

With these definitions of rms current and voltage, the average power equations become:

$$\begin{aligned}
\langle P \rangle &= \frac{1}{2} V_{peak} I_{peak} = V_{rms} I_{rms} \\
&= \frac{V_{rms}^2}{R} \\
&= I_{rms}^2 R
\end{aligned} \tag{7.6}$$

AC Voltage, Current and Power Examples

- Question: A hair dryer is plugged into the 120 VAC outlet, and consumes 1275 W of AC power. How much AC current goes through the hair dryer?

  Answer: The power of an AC device is the average power, unless otherwise noted. The average AC power is 1275 W, and the rms voltage is 120 VAC. $P = IV$ holds for these quantities, so the current is $1275/120 = 10.6$ Amps rms.

- Question: In the above problem, what is the resistance of the hair dryer?

  Answer: $V = IR$ holds for rms AC quantities. Therefore the resistance is V/I = 11.3 Ω.

- Question: In the above problem, what is the peak power?

  Answer: In an AC circuit, the peak power is just twice the average power: 2550 W peak power.

- Question: 'House current' refers to 120 VAC that comes from the wall outlets in your house. What is the mathematical expression for house current?

  Answer (part II): The general mathematical expression for AC voltages is $V_{peak} \sin(2\pi ft)$, and we have found above that the sine portion of the expression is $\sin(377t)$. The voltage is 120 VAC, meaning $V_{rms} = 120$. From the definition of V_{rms} above, $V_{peak} = \sqrt{2} \times 120 = 170$ V, so the expression for wall socket voltage is $170 \sin(377\,t)$.

In summary then: If current and voltage values are rms values, then calculating the average power or the resistance uses formulas that are the same as the familiar DC formulas. If peak values are important, convert between rms and peak values using Equations (7.4) and (7.5).

7.3 Current-Voltage in a Capacitor

In this section we discuss the relationship between current and voltage and develop the I-V plot for a capacitor in an AC circuit. We will find a ratio between V_{rms} and I_{rms} similar to Ohm's Law.

For a resistor, the V vs. I plot is a straight line, such as Figure 7.1. When an AC voltage is applied across a capacitor, the situation is quite different. If the voltage starts at V = 0 and increases from there, the *charge* on the capacitor starts at Q = 0 and increases rapidly, because Q = C*V at every instant in time. Therefore the current is already large when the voltage is just passing through zero. However, when the AC voltage reaches its peak value, the charge is peaked out, so I goes to zero at max voltage.

To explain this further, the time-dependence of charge, current and voltage appear in Figure 7.2, with time as the horizontal axis. The three dashed lines in the figure indicate points in time when the charge reaches a maximum or minimum, and the current has a zero.

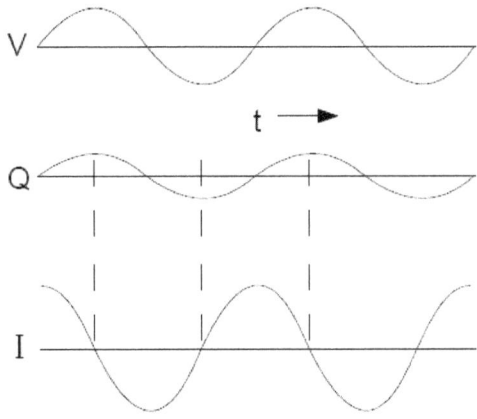

FIGURE 7.2
In a capacitor, the charge Q tracks the voltage, but the current is proportional to the slope of the charge curve.

If one traces the time dependences of Figure 7.2 onto I-V axes, the I-V plot for a capacitor is shown in Figure 7.3. As current and voltage vary with time, the I-V point moves counter-clockwise around an ellipse. At $t = 0+$, the voltage is zero and increasing, whereas the current is just falling off from its maximum value.

This is indicated by the black dot and arrow in the figure. This is quite different from the straight line I-V curve we obtained for a resistance!

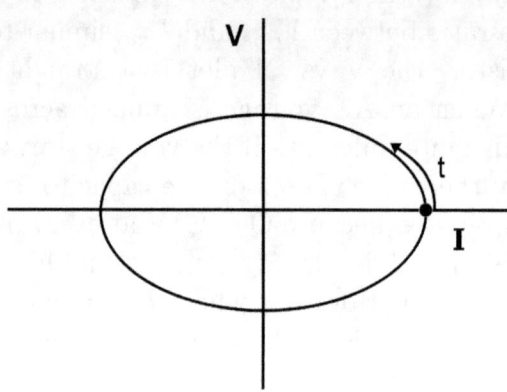

FIGURE 7.3 Parametric plot of I and V in a capacitor with time as a parameter. The I-V plot starts at $I = I_{max}$ and $V = 0$, and proceeds counter-clockwise around the ellipse.

Although Figure 7.3 does not contain a slope ratio between I and V, the dimensions of the ellipse do give a ratio that characterizes the capacitor's behavior. The vertical size is the applied voltage amplitude. The current (horizontal size) will be bigger if the capacitance is bigger, because a larger capacitor stores more charge and therefore takes more current to charge and discharge. The AC frequency is also a factor. At zero frequency (DC), there will be no current, and the ellipse shrinks to the y-axis. As frequency goes higher, the capacitor current will increase, and the ellipse widens. The current $I = \Delta Q/\Delta t$ gets bigger at higher frequency because the amount of charge is always the same ($= C*V$) but Δt is smaller. The exact result for the rms current is:

$$I_{rms} = V_{rms} * 2\pi f C \quad \text{where f is frequency, and thus}$$

$$X_C \equiv \frac{V_{rms}}{I_{rms}} = \frac{1}{2\pi f C} \quad \text{capacitive Reactance} \quad (7.7)$$

where X_C is called the *capacitive reactance*, defined as the ratio V_{rms}/I_{rms} in analogy with Ohm's Law. In a capacitive AC circuit, reactance takes the place of resistance in Ohm's Law. It provides the relationship between rms current and voltage. It is important

AC Circuits

to keep in mind, however, that at each instant in time the current and voltage in a capacitive circuit do not track together, as they do for a resistor. We say that they are 90° out of phase, with the current *leading* the voltage 90° because the current reaches it's peak $\frac{1}{4}$-cycle before the voltage reaches its peak.

7.4 Current-Voltage in an Inductor

Section 6.3 mentioned inductors and self-inductance in connection with Faraday's Law. A coil or solenoid creates an opposing emf when the source of the applied magnetic field is the coil itself. One can imagine this inductive response becomes quite significant in AC circuits, where the current is changing 100% of the time. This section develops the I-V plot of an inductor, and defines the inductive reactance, which is completely analogous to capacitive reactance and Ohm's Law for resistors.

The inductor is a coil of wire. It allows DC current to pass freely through it, because there is no Faraday effect when the current is steady. However, when the current is changing, a large voltage is induced. The time-dependence of voltage and current in an inductor is shown in Figure 7.4. The dashed lines indicate where the fastest changes in current result in maximum absolute voltage across the inductor.

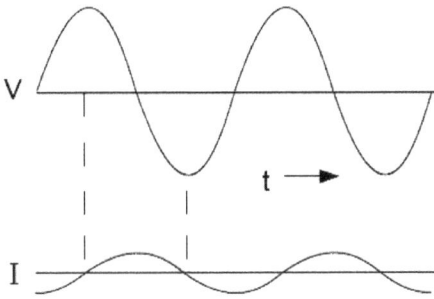

FIGURE 7.4
In an inductor, the slope of the current determines the voltage.

Tracing the time dependences of Figure 7.4 onto I-V axes, the I-V plot for an inductor is shown in Figure 7.5. As current and voltage vary with time, the I-V point now moves clockwise around the ellipse. This is similar to the capacitor ellipse plot, but the point moves in the opposite direction! Again, the dimensions of

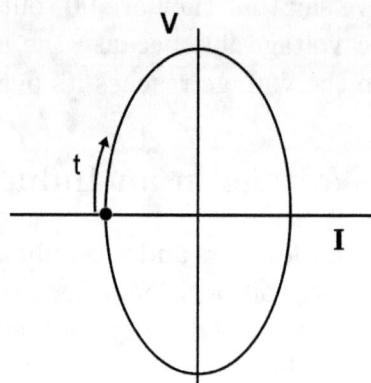

FIGURE 7.5 Parametric plot of I and V in an inductor with time as a parameter. The I-V plot starts at $I = I_{min}$ and $V = 0$, and proceeds clockwise around the ellipse.

the ellipse give a ratio that characterizes the inductor's behavior. The vertical size is again the applied voltage amplitude. The current (horizontal size) will be *smaller* if the inductance is bigger, because more turns on the inductor add up to a larger induced emf for a given current change. The AC frequency is again a factor. At zero frequency (DC), there will be no induced voltage, and the ellipse shrinks to the x-axis. As frequency goes higher, the current changes more rapidly, and less current flows for a given induced voltage. The ellipse becomes narrower and taller. The exact result for the rms induced voltage is:

$$V_{rms} = 2\pi f L I_{rms} \quad \text{where f is frequency, and thus}$$
$$X_L \equiv \frac{V_{rms}}{I_{rms}} = 2\pi f L \quad \text{inductive reactance} \tag{7.8}$$

where X_L is called the *inductive reactance*, defined as the ratio V_{rms}/I_{rms} in analogy with Ohm's Law. Equation (7.8) provides the relationship between rms current and voltage in the inductor. Note again, however, that at each instant in time the current and voltage in an inductive circuit do not track together, as they do for a resistor. Like the capacitor, they are 90° out of phase, with the current now *lagging* the voltage 90° because the current reaches it's peak $\frac{1}{4}$-cycle after the voltage reaches its peak.

7.5 Power in a Capacitor or Inductor

The I-V relationships for capacitors and inductors in Equations (7.7) and (7.8), resp., provide a substitute for Ohm's Law when these devices are used in AC circuits. An important question is what is the flow of power for these devices in AC circuits? This section shows that although energy is constantly flowing into and out of a capacitor, the average power is zero. The same holds true for an inductor.

Figure 7.6 shows the time dependence of instantaneous voltage, current and power when an AC voltage is applied to a capacitor. The + and − signs show the power spends equal amounts

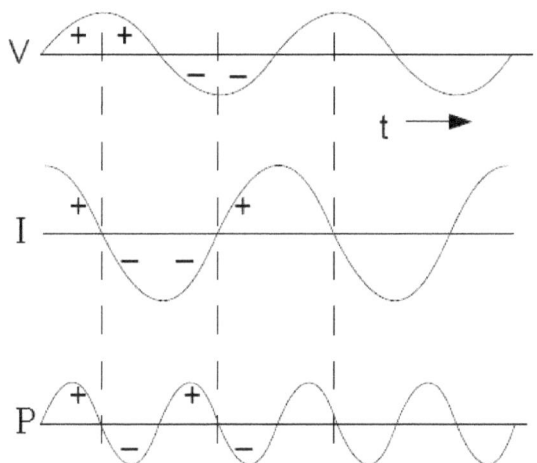

FIGURE 7.6 Instantaneous voltage, current and power in a capacitor. The power plot is the point-by-point product of voltage and current at each time t.

of time in positive and negative territory, and so adds to zero on average. One can also see this from trigonometric identities: The voltage is a sine function, the current is a cosine, and so the product is

$$P = V_p I_p \sin 2\pi ft \cdot \cos 2\pi ft = \frac{V_p I_p}{2} \sin 4\pi ft \qquad (7.9)$$

The power, $\sin 4\pi ft$, is a sine function, so has zero average value. The capacitor stores energy in its electric field when the voltage is building up and current is flowing into the capacitor. When the voltage turns around, the same amount of current flows out of the capacitor and drains the energy out of the electric field.

The same argument and similar graphs hold for the inductor. Current and voltage are again 90° out of phase, so the power averages to zero in an inductor. When the magnetic field is building, the voltage and current are positive and energy is flowing into the inductor. As soon as the current starts down, the voltage reverses and the current-voltage product goes negative, draining the energy out of the magnetic field.

In summary, reactors such as capacitors and inductors are similar to resistors in that the AC current is proportional to the AC voltage. However, they differ in that

1. The instantaneous current and voltage are out of phase in reactors

2. Reactors can store energy momentarily, but the average power is zero

3. The reactance of a capacitor varies inversely, and the reactance of an inductor is proportional, to the frequency

Circuits incorporating capacitors and inductors make great use of their frequency dependence and their energy storage capabilities.

7.6 Transformers

We have learned in Physics: Part I that enough energy by itself is not sufficient to accomplish jobs we need to do. It is also necessary for the energy source to match other aspects of the task, such as the required force, velocity, or mass into which the energy is being channeled. We showed that a variety of mechanical tools — e.g., the lever — transform the energy so that it can be supplied with the force or speed that the job demands.

Matching is also extremely important in electrical circuits. We saw in Sections 3.6.2, 3.6.3 that power sources can be interconnected so that their combined output voltage or internal resistance

AC Circuits

matches the load requirements. Often the source of energy has voltage that is too high for other devices, and we use voltage dividers, such as shown in Figure 3.9, to reduce the voltage to a safe range.

For AC circuits, there is a device, called a *transformer*, that is especially efficient in matching voltage and current requirements. We will see that transformers act like levers to change the voltage and current to fit a power source to the load, yet, like a lever, they dissipate very little of the power within themselves.

Figure 7.7 illustrates the basic principle of the transformer. The transformer consists of two coils wrapped around the same

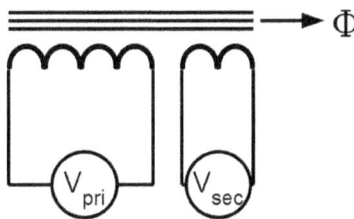

FIGURE 7.7
A transformer consists of two coils sharing the same magnetic flux.

magnetic flux path. These coils are called the *primary* and the *secondary*, as indicated in the figure. The three parallel lines in the figure represent a ferromagnetic core that enhances the linkage between primary and secondary by providing an easy path for the magnetic flux.

The primary and secondary coils are usually each a part of a separate circuit. Any change in the circuit of one coil affects the circuit containing the other coil. This is due to Faraday's Law, Chapter 6. Suppose we do something to change the current in the primary. Perhaps we apply a changing voltage. This produces a proportionately changing magnetic field generated by the primary, and therefore a proportionate change in the magnetic flux. But the transformer is designed so the primary and secondary coils both share the *same* magnetic flux. Therefore the secondary experiences a change in flux, inducing a voltage in the secondary.

The most common use for transformers is in AC circuits. Then the voltages and currents in the primary are changing continuously, and Faraday's Law implies that AC voltage is constantly being induced in the secondary. There is a fixed relationship between the size of the voltage in the primary and the secondary. It is derived as follows: for the primary and secondary voltages, V_p and V_s, Faraday's Law dictates that

$$V_p = -N_p \frac{\Delta \Phi}{\Delta t}$$
$$V_s = -N_s \frac{\Delta \Phi}{\Delta t}$$

and, dividing these two equations

$$\frac{V_s}{V_p} = \frac{N_s}{N_p} \tag{7.10}$$

where $\Delta \Phi$ is the change in flux, common to both primary and secondary; Δt is any particular time interval; and N_p and N_s are primary and secondary turns, respectively. Equation 7.10 gives the very important result for transformers, that

- The voltage ratio between primary and secondary in a transformer is in the same proportion as the turns ratio.

This makes the transformer very useful in changing AC voltage to suit particular applications. For example, we standardize most household and commercial appliances, lighting, etc., to operate at 110-120 VAC. This is a convenient voltage for household wiring. The reasons are that higher voltages present more of a shock hazard, whereas lower voltage requires more current, and therefore thicker wiring, to deliver enough power. However, wherever useful devices require higher or lower voltage, an appropriate transformer with the right turns ratio can provide power at the correct voltage.

In this regard the transformer is very analogous to the mechanical lever. Humans have the energy to lift an automobile, but human muscle power cannot provide enough force to lift a car bare-handed. Instead we use a mechanical jack. This lever is able to transmit energy from muscles to car with a much greater force. Or, on the contrary, we often want to apply a force that is smaller but at a higher velocity. A reverse lever, such as a fly

AC Circuits

swatter or tennis racket, permits very high speed, but a smaller force, appropriate to a light-weight object like a fly or a tennis ball.

Although a lever allows us to transmit power with a force or velocity to match the job at hand, the lever does not save any energy. Nor does it absorb any energy, assuming it is reasonably frictionless. The lever is power-neutral— i.e., we input a smaller force with the handle but push it farther, and thereby we lift the car a small amount with a much greater force. A simple lever is depicted in Figure 7.8. The product of force × distance is the same on both sides of the lever. The energy input on the right, $F_P \cdot \Delta y$, is equal to the energy output on the left, $F_W \cdot \Delta z$.

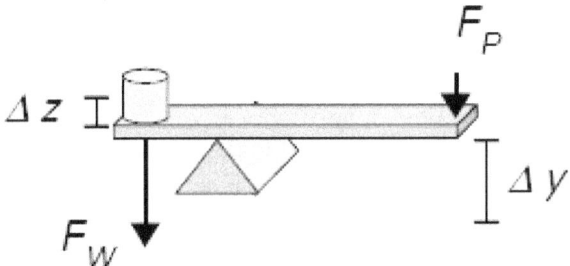

FIGURE 7.8
Like a transformer, the mechanical lever can transmit power with a greater or lesser force, chosen to match the job at hand. Energy is conserved.

We can extend the analogy between the transformer and the lever to the transmission of power. Just as mechanical power equals force × velocity, AC electrical power equals voltage × current. The transformer matches power to the job at hand either by raising the voltage— but at a reduced current; or by increasing the current— but at a reduced voltage. In this way the power going into the transformer equals the power coming out of the transformer. Losses such as resistance of the copper and 'friction' in the alternating motion of the magnetic domains can usually be ignored. The transformer itself neither creates nor dissipates energy. The mathematical equation that expresses this fact is

$$V_s I_s = V_p I_p \qquad \text{conservation of energy} \qquad (7.11)$$

- The power input to a transformer primary is equal to the power output from the secondary.

The conservation of energy, Equation (7.11) requires the current ratio between the transformer primary and secondary to be the inverse of the voltage ratio. Dividing (7.11) by $V_s \cdot I_p$, obtain

$$\frac{I_s}{I_p} = \frac{N_p}{N_s} \qquad (7.12)$$

- The current ratio between primary and secondary in a transformer is in the inverse proportion as the turns ratio.

7.7 Step-up and Step-down Transformers

A *step-up* transformer is used to multiply the voltage in the primary to produce a higher voltage. Typical applications of high voltage include cathode ray and gas discharge tubes, such as neon signs, ignition sparking devices, and electroshock weapons such as Tasers.

Step-up Transformer Example

- Question: A neon sign transformer steps-up the voltage from the power line (115 VAC) to 15,000 Volts. What is the turns ratio?

    Answer: The ratio is $15000/115 = 130$. Therefore if there are $N_p = 200$ turns in the primary coil, there must be $200 \times 130 = 26{,}000$ turns in the secondary coil.

Step-down transformers are used to produce low voltage or high current. Typical applications include low-voltage outdoor lighting, and high-current power supplies for welding. Many small electronic devices use low-voltage transformers which step down house line voltage in order to replace or recharge batteries.

Step-down Transformer Example

- Question: A model train has a power supply that plugs into the wall (117 VAC). The train runs with a maximum voltage of 15 Volts. At this voltage it draws 4.0 A. What is the power used by the train? What current is drawn by the primary?

  Answer: The ratio is $117/15 = 7.8$. The output power is 15 V $\times$ 4.0 A = 60 W. The primary current must be this same power divided by 117 V, or 60 W/117 V = 0.51 A.

FIGURE 7.9 Step-down transformers reduce the high voltage used for transmitting power over long distances to voltages proper for use in the home.

One of the most familiar and important applications of transformers is to step up electric generator output to very high voltages in order to transmit electrical power efficiently over long distances. It is then necessary also to step down the voltage at the destination for general use in factories and residences. See Figure 7.9 for the familiar transformer on top of telephone pole in a local neighborhood. The reason for transmitting power at high voltage is that the wires are limited in the amount of current they can carry; otherwise excessive heating and waste of energy would occur. With homes typically drawing hundreds of Amperes at 110 or 220 V under peak loads, a power line that is reasonable size, weight, and cost could only furnish a few dozen houses with the required amount of current. Therefore low voltage is used only locally for short distance runs, where the wire resistance will not be a factor. Long distances must be traversed by 'high tension' cables which deliver high power using low current and high voltage.

- Question: What is the turns ratio of a step-down transformer designed to convert 200,000 VAC to 120 VAC? If the transformer secondary supplies a neighborhood with a maximum of 10,000 A, how much current does this cause to flow in the primary?

  Answer: The ratio is 200,000/120 = 1667. The output power for the neighborhood is 120 V × 10,000 A = 1.2 MW. The primary current must be this same power divided by 200,000 V, or 1.2×10^6 W$/2 \times 10^5$ V = 6 A.

7.8 Power in the Secondary

The characterization of transformer voltages and currents in Section 7.6 gave the magnitudes, but did not completely describe the directions of the currents and voltages in the transformer. This section digs deeper into voltage and current directions, and how energy transfers through the transformer.

7.8.1 Transformer Delivering Low Power

Figure 7.10 shows the circuit of a step down transformer operating at zero power. The switch S is open, so no current flows in the secondary. A small current flows in the primary. Indeed, the transformer is designed to have very large magnetic flux for minimal flow of current— for example, the transformer uses many turns of wire and a soft iron core to enhance the magnetic field, both of which maximize the flux. Hence, in this example, it takes very little primary current— ≈ 0.6 rms Amps—to create a sizable flux. This flux operates jointly on the primary and the secondary. By Faraday's Law, it produces primary and secondary voltages in proportion to the turns ratio, as seen in Equation (7.10). Figure 7.10a shows the time-dependence of those voltages. The primary

AC Circuits

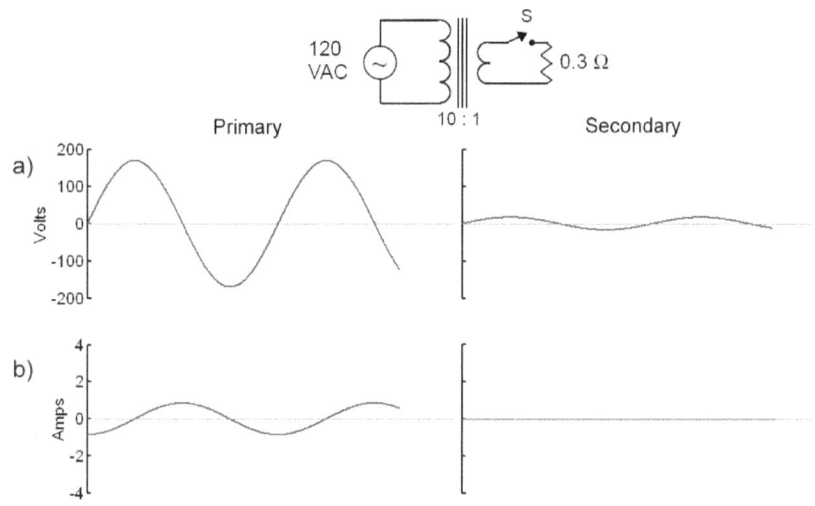

FIGURE 7.10
A 10:1 step-down transformer whose primary is connected to 120 VAC power and whose secondary is an open circuit. a) The voltage in the primary and secondary. b) The current in the primary and secondary.

voltage is 120 VAC, while the secondary voltage is 10× smaller, or 12 VAC.

Figure 7.10b shows the time-dependence of primary and secondary currents. As for all inductors— see Section 7.4— the primary current lags the voltage by 90°. Again, as for inductors, the average power in the primary is zero, because the product $I \cdot V$ spends equal time positive and negative. The power in the secondary is zero, because zero current flows.

Low Power Summary:

- When the transformer secondary is open-circuited or lightly loaded, there is very little primary current— just enough to induce the expected primary and secondary voltages. There is no flow of power through the transformer.

7.8.2 Transformer Delivering High Power

Figure 7.11 shows the same step-down transformer with switch S closed. Now the load takes significant power. The secondary voltage has produced a large current through the resistor, shown in

FIGURE 7.11 The same step-down transformer as Figure 7.10 now connected to a 0.3 Ω load. c) The secondary voltage causes a large secondary current, and a smaller out-of-phase primary current. d) The power produced by the primary is negative — i.e., power flows into the primary from the AC source.

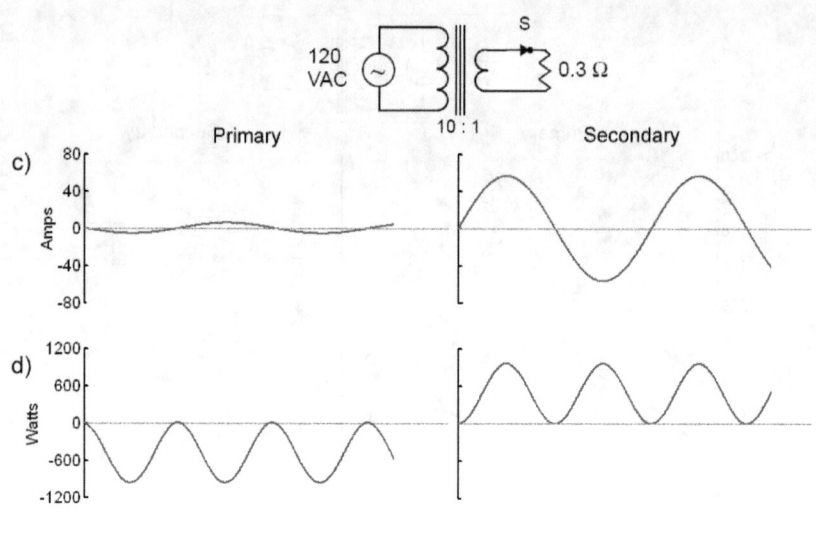

Figure 7.11c. Why are primary and secondary currents in nearly a 1:10 ratio and almost exactly *out of phase*? Here is the reason: by Faraday's Law the magnetic flux is directly proportional to the primary voltage— and the primary voltage is fixed by the power source. This limitation on the magnetic flux poses a limit on the *total* current in the transformer— i.e., the net current of the two coils. They are out of phase so that the sum of primary and secondary currents must also be limited. There can be large currents flowing, but they must add up to a small value. Hence they must be nearly opposite phase.

Restating this in different terms, the construction of the transformer (large number of turns, big coil area, high susceptibility iron core) is such as to respond to any net current with a big magnetic flux Φ. Therefore the primary current responds so as to almost completely cancel the flux generated by the big secondary current. Since the primary in this example has 10× as many turns as the secondary, it requires only one-tenth the current, 180° out-

AC Circuits

of-phase, to nullify the flux produced by the secondary current.

Specifically in terms of numbers, the secondary current is $I_s = 40$ A rms. The primary current needed to nearly cancel the flux is $I_p \approx -4$ A rms. Don't forget that we still need a small 90° out-of-phase primary current ≈ 0.6 A rms to provide the voltages, so I_p is not *exactly* -4 A rms.

Figure 7.11d shows the instantaneous power $I \cdot V$ in the transformer. The small net flux in the transformer produces a positive voltage having the same direction in both the primary and the secondary. However, large primary and secondary currents flow in opposite sense. A large amount of power enters the primary and leaves the secondary to drive the load. The conservation of energy Equation (7.11) is really a consequence of the transformer currents flowing in opposite sense to a $\approx$zero-sum net flux in the transformer.

High Power Summary:

- When the transformer secondary delivers high power to a load, the primary and secondary currents are nearly 180° out of phase, so as to limit the net flux in the transformer. Therefore power flows into the primary, and approximately equal power flows out of the secondary to the load, as per Equation (7.11).

Chapter 8

Waves

Objective: Wave motion has the utmost importance in our universe. Wave motion permits energy and information to transfer from one object to another, without actual contact between the source and the receiver. In the beginning, "the Lord said, 'Let there be Light'..." Thus the very first chapter of the Old Testament recognizes the significance of waves: (1) Light allows us to see; hence the transmission of information across indefinite distances. (2) Light brings the Sun's warmth and energy needed to produce all sustenance on Earth. The subject of waves includes not only electromagnetic waves such as light, but also mechanical and pressure waves such as sound. The discussion of waves leads to aspects of music as well as to optics and the workings of the human eye. It extends to modern insights like the wave-particle nature of matter and the structure of atoms.

In Physics Part I we studied the oscillation of a mass on a spring. In that system, energy oscillates back and forth between the potential energy of a stretched spring and the kinetic energy associated with the velocity of the mass. When any flexible medium such as air, water, a stretched trampoline, etc., is disturbed by moving it at one point, the kinetic energy input soon stores up locally as potential energy in a nearby location. As time progresses, that potential energy releases again into kinetic energy at a further location, etc. There results a succession of regions with kinetic and potential energy which march out together from the point of the disturbance at a specific speed in all possible directions.

We will see that all waves share the above characteristics. For electromagnetic waves, the initiating disturbance could be an electric current moving in an antenna. Waves always involve a transmission of energy in a succession of moving regions. For example, in sound waves, those regions are called 'compressions' and 'rarefactions', which are places of higher and lower than normal pressure, resp. In an electromagnetic wave, the energy is contained in moving regions of high electric and magnetic field. The disturbance moves through space with the speed of light in the case of electromagnetic waves in free space; and at the speed of sound in the case of pressure waves in an elastic medium such as air.

8.1 Waves on a Rope

If you tie one end of a long rope to a post or tree, put some tension on the free end with one hand, and pluck the rope with the other hand, a wave step will propagate toward the tree, reflect back to your hand, and bounce back and forth several times before the energy in the wave dissipates. Figure 8.1a shows a slanting step traveling on the rope, and 8.1b shows the same piece of rope a short time later. In the time interval, the step has moved a little to the right.

In Figure 8.1a, the tension in the rope is creating an unbalanced upward force on the segment highlighted with a (*). That is because on the left end of the segment, the rope tension is pulling up and to the left. On the right end of segment (*), however, the rope tension is pulling only horizontally to the right. The net force accelerates the segment so that it gains an upward velocity. Because of its new upward location at time Δt later in Figure 8.1b, segment (*) is ready to pull upward on the next segment

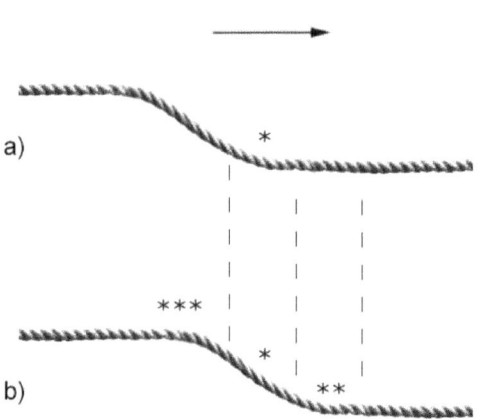

FIGURE 8.1
(a) Wave-step traveling to the right on a rope. (b) Same wave, a short time Δt later.

(**) to the right. This permits the wave to propagate forward. Momentum keeps the slanting segment going upward until it encounters the downbend in the rope at location (***). When it passes through the downbend, the net force on the segment (*) will become downward, stop its rising, and bring it to rest at the new height. This part of the rope will be stationary again, and the wave will have passed by.

For this wave the propagation depended on the rope tension to accelerate the piece of rope through which the wave was passing. The mass of the rope presents inertia that limits the acceleration. It should not be surprising, then, that the wave speed, c, for waves on a rope or string is given by

$$c = \sqrt{\frac{T}{\sigma}} \tag{8.1}$$

where T is the force of tension in the rope and σ is the mass per unit length. Although Figure 8.1 showed the propagation of a sloping step, any shape of disturbance will propagate along the rope, with essentially no change in shape, at the wavespeed given by Equation (8.1). We will see that every medium transmits waves of several types, and in each case the motion has a characteristic speed that depends on various properties of the medium.

8.1.1 Frequency, Wavelength, and Wave Speed

The above Section 8.1 discussed a step propagating on a rope. Instead of steps or pulses, often an *oscillating* disturbance creates the wave. This generates a repetitive motion that propagates as a sine wave through the medium. This is called a *continuous wave*, as opposed to a step or pulse. Sinusoidal wave motion is common to many forms of energy transfer, such as sound, water waves, radio, light, x-rays (i.e., all electromagnetic waves). Figure 8.2a is a snapshot of a continuous sinusoidal wave propagating to the right. The figure illustrates some important terminology. If we

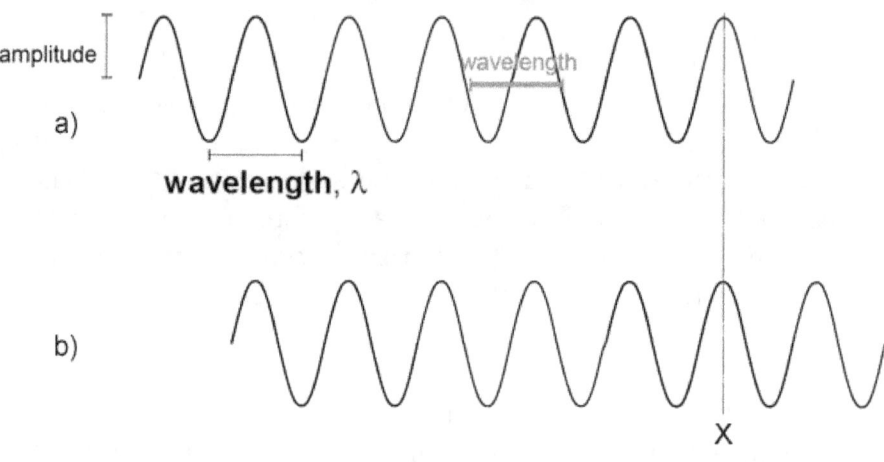

FIGURE 8.2 (a) Snapshot of a sinusoidal wave at a fixed point in time. The distance between successive peaks is the *wavelength*. (b) The same wave viewed at a time one *period* later. The wave has moved one wavelength to the right during this time interval.

stand at a single point in space, say point X, and watch the waves go by, the motion is sinusoidal in time. The time interval between successive peaks, or between successive valleys, is one *period*. The typical symbol of the period is **T**, and the unit is the second.

Figure 8.2b shows a snapshot of the same wave one period later. The wave has shifted in position by a complete $360°$, or 2π radians. The space occupied by one complete cycle of the sine wave is called the *wavelength*. The unit of wavelength is the **meter**, and the usual symbol is λ, Greek letter lambda.

The wave has progressed a distance λ in time **T**. Therefore the wavespeed is

$$c = \frac{\lambda}{T} \qquad (8.2)$$

where c is the wavespeed of the wave, and has units m/s. For light the wavespeed in free space is a well-known constant of nature, $c = 3.0 \times 10^8$ m/s. The speed of sound depends upon the air temperature and pressure, but a typical value is $c_{sound} = 330$ m/s.

Frequency

A way of counting the rate of oscillations in a wave is to consider how many complete cycles pass a fixed point in space in one second. This is called the frequency, f, of the wave. Since a wavelength passes a fixed observation point in T seconds, the number of waves per second is 1/T:

$$f = \frac{1}{T} \quad \text{and, substituting in Eq. (8.2),} \qquad (8.3)$$
$$c = f\lambda \qquad (8.4)$$

8.1.2 Interference

What happens when two waves traveling in opposite directions collide? If you and a friend face one another and talk at the same time, do your words interfere with your friend's words? Aside from the obvious difficulty of talking and listening at the same time, the answer is no. This is quite unlike communicating by throwing stones at each other. Stones can certainly collide and be deflected. The two sound waves, however, pass right through each other without any permanent change occurring in either wave. Similarly, the beams of two flashlights do not affect one another, and the same can be said of water waves traveling at an angle to each other. However, the water will be significantly disturbed or choppy at the point where the two waves cross paths. At the point of waves crossing, the water, or other medium, moves a total amount which is the simple sum of the two waves at that time.

The fact that waves pass right through one another unchanged, but affect the medium they pass through in an additive way, is called *interference*. If two waves are deflecting in the same direction at a moment in time at a particular point in space, this is called *constructive* interference. On the other hand, if the two waves have opposite deflections at that point, the algebraic sum can be smaller than either one of the two individual waves. This is *destructive* interference.

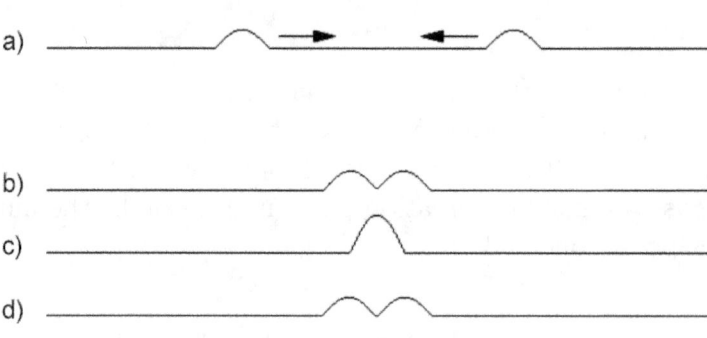

FIGURE 8.3 a) Two pulses travelling in opposite directions. b)-e) Snapshots at consecutive later times. c) Constructive interference.

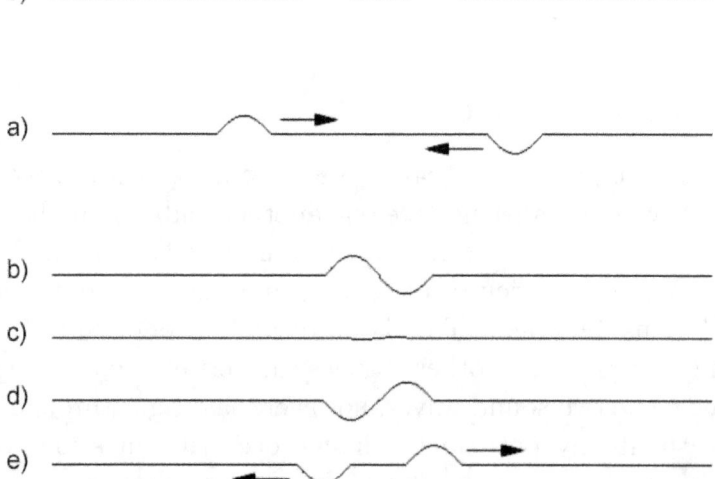

FIGURE 8.4 a)-e) Snapshots of two pulses travelling in opposite directions. c) Destructive interference.

Figure 8.3a-e shows the effect of two pulses interfering constructively, while Figure 8.4a-e shows two pulses with destructive interference. A more complicated but important case is when two continuous waves, traveling in opposite directions, interfere. This is depicted in Figures 8.5a-h. The two wave trains cancel, add,

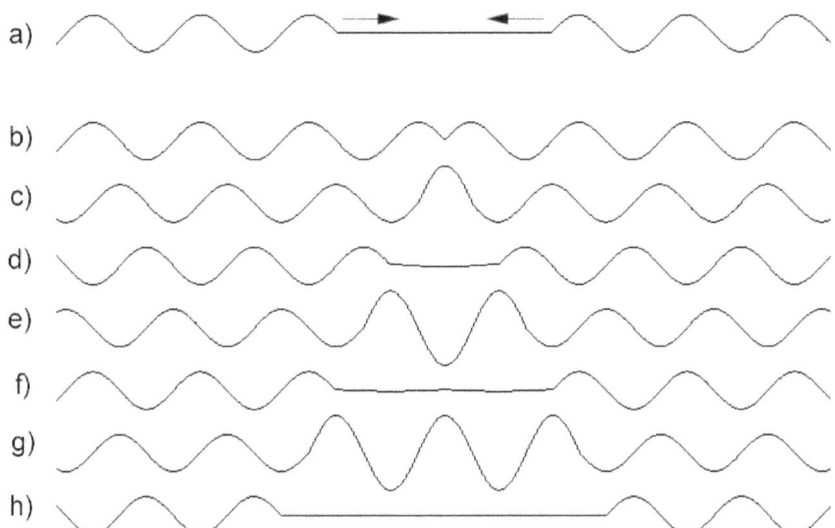

FIGURE 8.5
Sequence of snapshots showing the collision of two oppositely moving wave trains. Note how the pattern of oscillations near the center of the rope becomes stationary.

cancel, add ...etc., during successive moments. Sinusoidal disturbances to the left and to the right of the picture start generating two sinusoidal wave trains, proceeding toward each other from opposite directions in Figure 8.5a. Figures 8.5b-h show how the two wave trains cross and interfere at time intervals of $T/4$, one quarter period.

In Figures 8.5b-c, the first peak of the left wavetrain interferes constructively with the first peak of the right wavetrain. At d, f, and h, the peak of one wavetrain exactly cancels the valley of the other wavetrain, causing destructive interference. At e and g, the peaks of the two wavetrains interfere constructively again.

The sequence of repeating constructive, destructive, constructive, destructive ... interferences within the overlapping region of the colliding wavetrains causes the medium to vibrate at the frequency of the individual waves. The amplitude of vibration varies, being maximum at some locations and *zero* at others. The pattern is an envelope of motion that stands still in space! This sounds complicated, but it will become simpler when we discuss *standing waves* below in Section 8.1.4.

8.1.3 Wave Reflections

When a wave on a rope encounters an end tied to a fixed post, the wave reflects back. You may have noticed that the reflected wave deflects in the opposite direction from the incident wave. That is, if a positive pulse on the rope approaches a firm boundary, the reflected pulse is a negative pulse. The sequence of events is depicted in Figure 8.6. In Figure 8.6a, a positive pulse is traveling

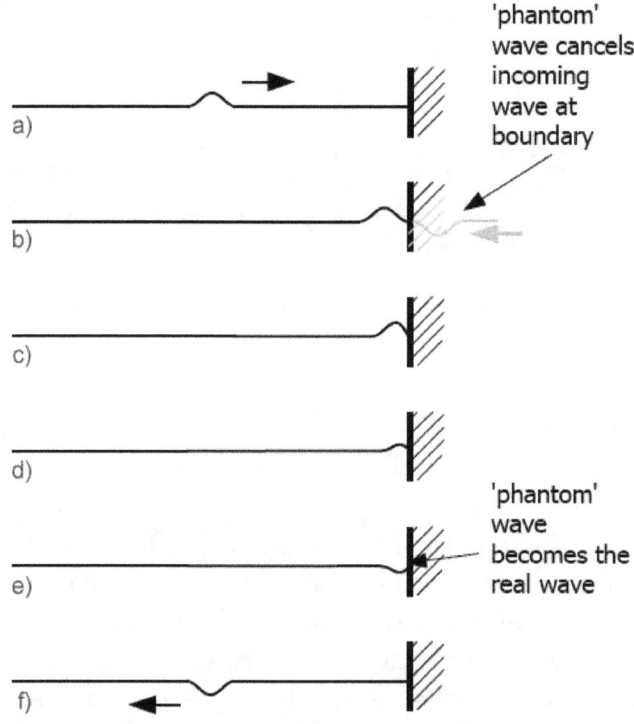

FIGURE 8.6 (a)–(f) Wave approaches wall and reflects, causing an inversion of the wave shape. See text for details.

on a taut rope toward a wall.

We can predict the shape of the reflecting wave by examining two waves traveling in opposite directions on a longer rope, with no wall. Figure 8.6b shows the two waves beginning to interfere. Because these two waves have opposite deflections, if they are timed right, they will exactly cancel when they meet at the location of the wall. Interposing the wall again, it would not constrain or distort the two waves at all. Figure 8.6c-e shows the further progress of the original wave and the 'phantom' wave of 8.6b interfering as they pass through one another. In Figure 8.6f, the phantom wave has emerged unaltered, and the upward pulse is out of the picture to the right. The model of two opposing waves exactly satisfies the boundary condition imposed by the wall. It correctly predicts the effect of a wave reflecting off a wall.

One reason the wave changes deflection when it reflects from a hard boundary is that the upward pulse on the rope, combined with the tension in the rope, pulls up on the wall. Recalling the discussion of Figure 8.1, the tension of the rope helps propagate the pulse by pulling up on the next segment of rope in the path of the wave. When it reaches the wall, it *tries* to pull up on the wall. At this point, the rope has encountered an immovable obstacle. By Newton's 3rd law, the wall pulls down with an equal and opposite force on the rope. Because the wall is firmly planted, the wave actually pulls itself into a negative deflection, rather than moving the wall upward.

8.1.4 Standing Waves and Resonance

Consider an elastic material such as a violin string or a slinky under tension. Again as in the previous section 8.1.3, the slinky is rigidly affixed at one end to a solid wall, and you launch upward pulses along the slinky toward the wall. The pulses reflect at the rigid boundary into downward pulses. When a downward pulse returns to you, it normally reflects into an upward pulse if you hold your end fixed. The pulse could continue to echo back and forth, eventually losing energy and dying out.

However, you may— at the precise moment it returns to you— give the slinky a slight additional upward jerk. This strengthens the wave pulse as it passes through your end, increasing its amplitude and energy. If you continue to provide this small assist every time the wave comes back to you, then pretty soon the whole length of the slinky is oscillating up and down with a large amplitude. This is called a standing wave. This build-up of this oscillatory motion is indicated in Figure 8.7a-f.

The initial reflecting pulse as it started out is shown in Figure 8.7a-b. The energy clearly is localize and transmits from one support to the other support, and back. In Figure 8.7c-d, the energy has spread out and the arrow indicates some side to side motion; however, the kinetic energy is becoming more centered in the middle portion of the slinky. Finally in Figure 8.7e-f, the motion of the slinky is entirely an up and down oscillation, with no side-to-side motion.

FIGURE 8.7 Development of a standing wave resonance. The distance between the rigid supports is L meters.

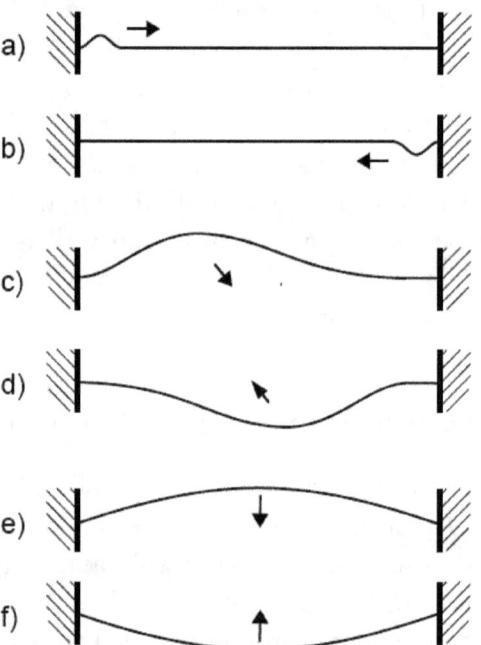

The period of this oscillation remains the same throughout Figure 8.7a-f. The time for the pulse in 8.7a-b to go the distance $2 \times L$, i.e., twice the length of the slinky, is the same as the time for the slinky to oscillate up and down as a unit in 8.7e-f. That period is $T = 2 \times L/c$ sec.

The ability for a wave-medium to absorb energy at a specific frequency is called a *resonance*. The frequency $f = 1/T = c/2L$

Hz is called the *resonant frequency*. Since there are many possible resonant frequencies, and this is the lowest, it is called the *fundamental* resonance frequency. When excited at the fundamental frequency, the slinky takes on a sinusoidal shape, with exactly π radians between two zeros of the sine function occupying the entire length L of the slinky. When excited at *higher harmonics*, the shape contains more peaks and valleys of the sine wave. For example, the second, third, and fourth harmonics are shown in Figure 8.8a-c.

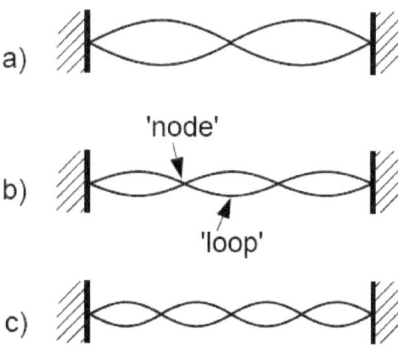

FIGURE 8.8
Second, third and fourth harmonics of the resonance shown in Figure 8.7e-f

The higher harmonics have higher resonant frequencies. The second harmonic, Figure 8.8, contains a full wavelength, or 2π radians of the sine wave, between the supports. Because an entire wavelength fits within the space L, There is a *node* or zero in the center, and two *loops* or *antinodes*— positions of maximum oscillation amplitude. There is also a node at either end, forced to zero movement by the rigid support. The following generalization always holds:

The shortest distance between two nodes (or two loops) is $\lambda/2$.

Parameters of the first several harmonics are shown in Table 8.1. There are an infinite number of possible harmonics. The pattern for the taut rope— but not necessarily the pattern for other resonators— is that harmonics start with the fundamental

Harmonic	No. of Nodes incl. Supports	Wavelength	Frequency
1	2	$2 \cdot L$	$c/(2L)$
2	3	L	$2c/(2L)$
3	4	$(2/3) \cdot L$	$3c/(2L)$
4	5	$L/2$	$4c/(2L)$
5	6	$(2/5) \cdot L$	$5c/(2L)$

TABLE 8.1: First five harmonics in the series of slinky resonances.

frequency, $c/(2L)$, and increase by multiples of the fundamental frequency. Each higher harmonic has an additional node and antinode.

8.1.5 Modes of Wave Transmission

You tie one end of a rope to a fixed support, and create waves by oscillating the free end. You can create waves in the up-and-down direction; also in the side-to-side direction. Both of these waves are called *transverse* modes, which means the motion of the rope is perpendicular to the direction of travel of the wave. The up-down and side-to-side directions are called vertical and horizontal *polarizations*. You can also move the free end in a clockwise or counterclockwise circular motion to create right and left *circular* polarizations.

Furthermore, the motion of your hand can be fast or slow, and complicated enough to produce an intricate, nearly random jiggling pattern along the rope. However, any transverse motion you impart to the rope can be broken down into the sum of specific proportions of the above harmonics and polarizations.

Transverse waves are not the only possible mode on the stretched rope. The rope can also oscillate forward and back along the direction of the rope. Such waves are called *longitudinal* waves. The longitudinal mode is harder to see on a taut rope, but a slinky makes it easy to visualize the longitudinal mode. Lay the slinky out on a table so it is slightly stretched. Then if you compress— then release— a section of the slinky, the compressed region propagates down the length of the slinky. In bygone days children used this longitudinal mode as a communication link between two paper cups. The paper cup funneled one child's voice into the string. The string conveyed the sound as a longitudinal wave to the other cup. The vibrations were directed out the other cup in to the second child's ear. In this way, two children were able to carry on a private conversation in the days before texting. The next few sections discuss sound in detail. All sound waves are longitudinal waves.

8.2 Sound Resonances

Sound resonances are similar to the resonances on a rope. Musical instruments are designed to produce resonances. We detect the accumulation of energy in such a resonance as a *note*. The frequency of the resonance corresponds to the *pitch* of the musical note. Higher frequencies give higher musical pitches. In fact if the frequency of a musical note is doubled, we hear the pitch of the note go up by one *octave*. Musical *chords* are mixtures of several pitches, whose frequencies are mathematically related, in a like manner as the frequencies of the resonances on a taut rope are mathematically related in 8.1.

The resonances in a pipe or on a string contain nodes and loops and we detect the resonances as notes with specific pitches. The resonances of a stringed instrument are the same as those of a taut rope discussed in Section 8.1. The resonances in a pipe, such as a flute, trumpet, or organ, are somewhat different because the boundary conditions— i.e., the constraints imposed by the ends of the pipe— are different from the constraints imposed by the endpoints of the rope.

Figure 8.9 shows resonances in two pipes. In 8.9a, a pipe closed at one end restricts any motion of air at the closed end, but allows free longitudinal flow of the air at the open end. On the other hand, compression and rarefaction— indicated by the

Harmonic	No. of Nodes inside Tube	Wavelength	Frequency
1	1	$4 \cdot L$	$c/(4L)$
2	2	$(4/3) \cdot L$	$3c/(4L)$
3	3	$(4/5) \cdot L$	$5c/(4L)$
4	4	$(4/7) \cdot L$	$7c/(4L)$
5	5	$(4/9) \cdot L$	$9c/(4L)$

TABLE 8.2: First five harmonics in the series closed-end tube resonances.

shading— can build up at the closed end. The open end cannot support any significant pressure build-up. Just below the pipe Figure 8.9a shows a quarter wavelength of the analogous transverse wave. The fundamental mode of an instrument is its lowest possible resonant frequency. Therefore, for a closed-end tube, the fundamental contains a quarter wavelength inside the tube. This is rather different from transverse waves on a string, where the fundamental mode contained one-half wavelength between the supports. Instead of Table 8.1, we have a new pattern for the harmonic frequencies in a closed-end tube, as shown in Table 8.2.

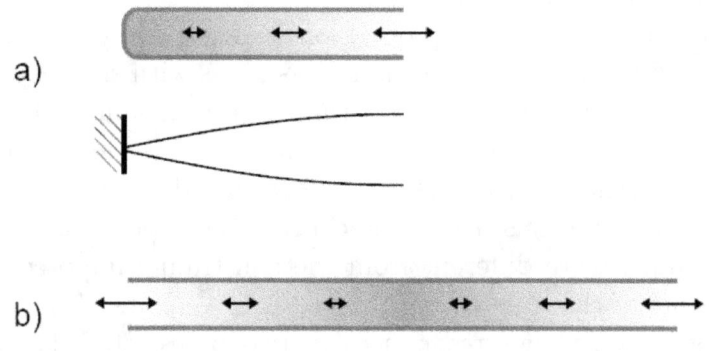

FIGURE 8.9
a) Air resonance in a tube closed at one end. Arrows indicate motion of the air. Also shown is the analogous wave on a rope. b) Resonance in a tube open at both ends.

The pipe open at both ends, Figure 8.9b, allows free flow of air at both ends. The longitudinal motion of the air has antinodes at either end of the pipe, and a node at the center. On the contrary, the pressure fluctuations have an antinode at the center of the pipe, and nodes at either end. The pressure amplitude in an open pipe therefore looks similar to the transverse amplitude on a rope or string, and Table 8.1 applies.

8.3 Electromagnetic Waves

Chapters 4 and 5 introduced the electric and magnetic force fields, and the forces they exert on charges and currents, respectively. Chapter 6 on Faraday's Law showed that when the magnetic field changes through a conducting loop, this generates an electric field. Therefore energy is associated with electromagnetic fields. And in dynamic situations where electric and magnetic fields change, even in free space without conductors present, the magnetic and electric fields become interlinked— i.e., changes in magnetic field cause the spontaneous appearance of an electric field, and *vice versa*.

When we generate longitudinal waves on a slinky— say, by pinching and releasing it— we create compressed and stretched regions in the medium which store energy. At first this energy is potential energy associated with the compression and stretching of a spring. However, the forces associated with compression and stretching cause acceleration of the medium. Parts of the slinky move and pick up kinetic energy. The disruption of the slinky propagates along the slinky as a wave containing both kinetic and potential energy.

The situation for generating electromagnetic waves is entirely analogous. If we create a local B-field by suddenly passing an electric current through a wire, there is energy associated with the magnetic field around the wire. The local change in the magnetic field creates an electric field in the immediate neighborhood, the changing electric field creates a magnetic field ... etc., and the resulting electromagnetic fields and their energy propagate away from the source in all directions at the speed of light, c. Electromagnetic waves are always transverse, meaning the electric field and magnetic field vectors are perpendicular to the direction of travel. A typical electromagnetic wave is depicted in Figure 8.10a.

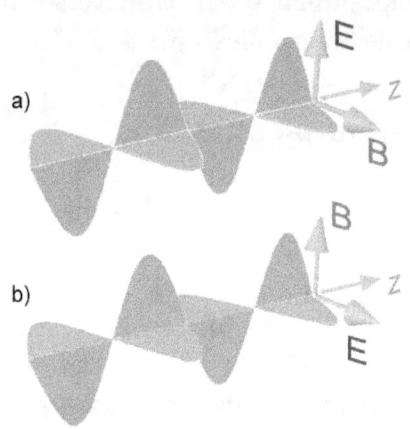

FIGURE 8.10 a) Vertically, and b) horizontally polarized electromagnetic waves traveling in the z-direction.

Figure 8.10a shows a snapshot of an electromagnetic wave moving in the z-direction. The electric field is oriented vertically. The electric field is varying sinusoidally as z changes, so it points up at one point in space, and then one-half wavelength further in the z-direction it points downward. In empty space there are no charges, but if the wave encounters a vertical conducting wire, the electrons in the wire would be jiggled up and down by the electric field. Similarly the B-field is varying horizontally, pointing first toward the reader and then away. If this wave passes a conducting loop in the plane of the paper, the B-field changes would cause current to flow clockwise and counterclockwise— due to Faraday's Law. Clearly it will be possible for this wave to transmit energy to a wire antenna or to a conducting loop antenna. By pulsing the electromagnetic wave, information can also be sent to such antennas.

Figure 8.10b shows a similar wave but with the electric field pointing horizontally. The magnetic field is now vertical. The wave in Figure 8.10a is said to be *polarized* vertically whereas the wave in Figure 8.10b is horizontally polarized. In other words, the direction of polarization is just the direction of the E-field. The situation is similar to the transverse waves on a rope, where the rope can wiggle up and down or in the side-to-side direction. The two E-field polarizations can also combine in arbitrary proportions to give waves polarized at other angles. If the horizontal and vertical components are shifted relative to each other in the z-direction, the resulting wave will be *circularly* or *elliptically* polarized. Circular polarization is entirely analogous to the circular motion which kids impart to a jump rope or you could give to a slinky, instead of the strictly up and down motion shown in figures such as 8.7e-f. We will return to some interesting aspects of light polarization in Section 9.3

8.3.1 The Electromagnetic Spectrum

In the previous Section 8.3, we mentioned how simple wires and loops can act as antennas to receive energy and information from electromagnetic waves. The use of antennas is familiar for television and radio, cell phones, and other techy wireless devices. It will be shown in Section 9.6 that the antennas are usually of a size comparable to the wavelength being received in order to gather the energy efficiently. For example, the frequency used for cellphones is 1.9 GHz. What then is the wavelength of the electromagnetic waves used for cellphones? Using the fact that the wavespeed for all electromagnetic waves is $c = 3 \times 10^8$ m/s, and Equation (8.2), we find that wavelength $\lambda = c/f = 3 \times 10^8 / 1.9 \times 10^9 = 0.16$ m, or about 6 inches. Hence a good antenna for cell phones should be at least several centimeters.

By comparison, FM radio has a typical frequency of 100 MHz, so the wavelength is $\lambda = c/f = 3 \times 10^8 / 10^8 = 3$ m. So a good FM radio antenna is considerably bigger than a cellphone antenna. My cellphone has a built-in FM radio, and to listen to music you have to plug in the earbud, because the wire that connects the earbud also serves as the FM antenna.

The reason wavelength is special for electromagnetic waves is the extraordinarily wide range of wavelengths (and frequencies) possible for this type of waves. AM radio waves, at a frequency of 1 MHz, have a wavelength of 30 meters. The waves used for global positioning systems (GPS) have a wavelength of about 10 m. Atoms are good antennas for x-rays, which have a wavelength of 10^{-10} m. Atomic nuclei are much smaller than complete atoms. They give rise to gamma rays, which are much more penetrating than x-rays and have even smaller wavelengths— 10^{-12} m. In principle, all wavelengths are possible for electromagnetic waves. Very important wavelengths in the evolution of plants and animals are the visible wavelengths ('Let there be light'). Deep blue or violet light has a wavelength of about 4×10^{-7} m. Red light has a wavelength up to about 7×10^{-7} m. Hence the visible light spectrum spans the range 400–700 nm. Ultraviolet has a little bit shorter wavelength than visible, and infrared has a little longer wavelength than visible light.

8.3.2 Energy Content of an Electromagnetic Wave

The warmth of the sun or hot coals, the beneficial and sometimes detrimental chemical changes produced by ultraviolet light, and the ability of x-rays to expose photographic film or digital sensors are all indications that electromagnetic waves carry energy. We have seen in Figure 8.10 that these waves contain electric and magnetic fields that move through space at the speed of light, c. These fields contain energy, as explained below. The movement of the wave carries the field energy at speed c as well.

E-field Energy

A good way to understand the energy of the E-field is to consider again the parallel plate capacitor. As depicted in Figure 4.7, the capacitor holds equal and opposite charges on the metal surfaces facing each other, and a constant E-field in between the two plates. There is an electric field in between the plates. Because the charges are opposite, they attract. Therefore it requires a force to hold the plates apart at a fixed distance, or to pull the plates apart. However, if we increase the plate separation against that force, we perform work, so the energy of the capacitor must increase. Where does the energy go? Assume no circuit connection to the capacitor, so there is no flow of current possible to use up the energy as heat. One reasonable place to look for the energy is in the E-field itself.

Following the application of Gauss' Law to capacitors in Section 4.5.3, a fixed charge on the plates implies the E-field is constant regardless of the plate separation. As we add energy by pulling the plates apart, the E-field does not change, but the volume occupied by the E-field increases in proportion to the distance d between the plates. Thus the E-field creates an energy density, and the energy in the capacitor is found by multiplying the energy density by the volume it occupies. This section explains why the energy density in a locale is given by:

$$U = \frac{E^2}{8\pi k} \qquad \text{energy density of field } E \qquad (8.5)$$

Here U is the energy density, E is the magnitude of the E-field, and k is the familiar constant 9×10^9.

Consider Figure 8.11, showing force F stretching apart the plates of a charged capacitor. The figure also shows a dashed

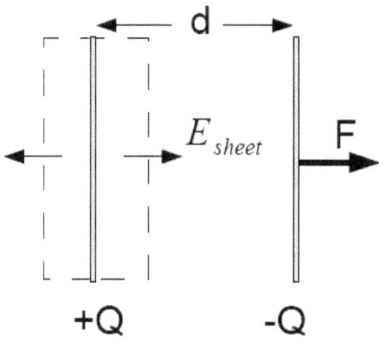

FIGURE 8.11 E-field energy calculation in a charged capacitor.

line indicating an imaginary volume for applying Gauss's Law.

The purpose is to find the E-field due to the (+)charged sheet on the left, and use this value of E-field to calculate the force on the (−)charged sheet on the right. We wouldn't want to count the entire E-field between the plates to calculate the force on the righthand plate, because that would include the force of the E-field from the righthand plate *on itself*. From Gauss's Law Equation (4.8), then, find that

$$
\begin{aligned}
2AE_{sheet} &= 4\pi kQ \\
E_{sheet} &= \frac{2\pi kQ}{A} \\
&= \frac{E}{2}
\end{aligned}
\qquad (8.6)
$$

where E in (8.6) is just the total E-field in the capacitor due to *both* plates. Therefore if the force $F = QE_{sheet}$ pulls the capacitor plate from separation ≈ 0 out to separation $= d$, the work input to the capacitor is

$$
\begin{aligned}
W &= Fd \\
&= QE_{sheet}d & \text{substituting for Q} \\
&= \frac{AE_{sheet}^2 d}{2\pi k} & \text{substituting for } E_{sheet} \\
&= \frac{E^2}{8\pi k} Ad
\end{aligned}
\qquad (8.7)
$$

Since the volume occupied by the E-field is $A \cdot d$, (8.7) means that the energy density is

$$
U_E = \frac{E^2}{8\pi k} \qquad \text{energy density of field } E \qquad (8.8)
$$

As an alternative way to demonstrate the validity of Equation (8.8), we will now show that the energy stored in a charged capacitor is just the energy stored in the E-field, as given by Equation (8.8). We previously derived in Equation (4.13) that the voltage difference V between capacitor corresponds to a stored energy,

Waves

$Energy = \frac{1}{2}CV^2$. C depends on the geometry of the capacitor through Equation (4.11). Substituting this value of C,

$$\begin{aligned}
Energy &= \frac{1}{2}CV^2 \quad &\text{and using Volume} = A*d, \\
Energy/Volume &= \frac{1}{2}\frac{AV^2}{4\pi kd}\frac{1}{A\cdot d} \quad &\text{and simplifying,} \\
U_E \equiv Energy/Volume &= \frac{1}{8\pi k}\left(\frac{V}{d}\right)^2 \quad &\text{then using } |E| = V/d, \\
U_E &= \frac{E^2}{8\pi k}
\end{aligned}$$

Therefore two different examples lead to the same expression for the energy contained in an E-field fixed in space between capacitor plates. We want to apply Equation (8.8) to find the relationship between the E-field and the power in the beam of an electromagnetic wave. The other half of the puzzle is the energy contributed by the B-field which necessarily accompanies the E-field in the wave.

8.3.3 B-field Energy

We saw that two parallel wires carrying oppositely directed currents repel each other. Unless the repulsive force is countered by an external force, the wires will move apart. The same is true for two current-carrying planar sheets, as depicted in Figure 8.12. In the figure, the left side is a current sheet with current coming out of the page toward the reader. The right side is an equal and opposite current going into the page. In Section 8.3.2 we evaluated the energy needed to pull apart the plates of a capacitor. We can also evaluate the energy needed to push the two current sheets together.

We can use Ampere's Law, Section 5.5, to evaluate the B-field. In Figure 8.12, the dotted line shows a path for applying Ampere's Law. We find that the B-field is entirely contained between the two current sheets, and is zero outside. It requires work to quickly push the sheets a little closer together. The number of

FIGURE 8.12 Diagram used to calculate the energy stored in a B-field. Oppositely directed current sheets confine a B-Field between them. Force F compresses the B-field. Equation (8.9) gives the B-field energy density.

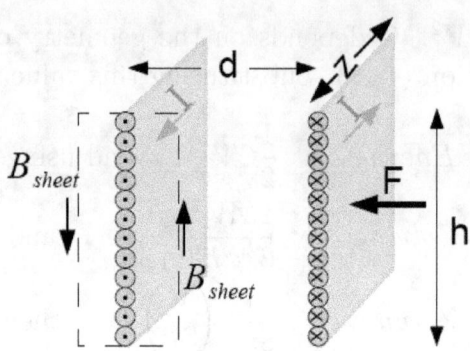

B-lines remains constant— they just become compressed between the current sheets into a smaller volume. This increases the value of the B-field between the plates.

On the other hand, the volume between the plates, $d \times h \times z$ in Figure 8.12, is reduced by pushing the plates closer. If we take all the energies into account, namely (1) the energy increase due to compression and (2) the energy decrease due to volume decrease, and (3) the work performed against the repulsive force between the plates, we can find an expression for the energy density due to the B-field:

$$U_B = \frac{B^2}{2\mu_0} \qquad \text{energy density of magnetic field } B \qquad (8.9)$$

Here U_B is the energy density, B is the magnetic field, and μ_0 is the familiar constant, equal to $4\pi \times 10^{-7}$.

Energy Transmission Examples

The energy carried by electromagnetic waves is the sum of the electric and the magnetic field energies, whose density in space is given by Equations (8.8) and (8.9). Because of rules, such as

Faraday's Law, relating E-field and B-field in an electromagnetic wave, these two energy densities turn out each to be equal halves of the total energy. Therefore the total energy density of the wave is

$$U_{EM} = U_B + U_E = 2U_B = 2U_E$$
$$= \frac{B^2}{\mu_0} = \frac{E^2}{4\pi k} \tag{8.10}$$

Equations (8.10) refer to the instantaneous values of energy density at each point in space through which the wave is traveling. One is probably more interested in how much power reaches an antenna, or how much radiant energy reaches a square meter of Earth from the Sun every second. To reach this goal, there are two concepts to apply:

1. The electromagnetic wave is sinusoidal, so there are points of the sine wave that have maximum energy density, and intervening points that have zero energy density. The average energy density is 1/2 the maximum:

$$U_{EMavg} = \frac{U_{EMmax}}{2} = \frac{E_{max}^2}{8\pi k} \tag{8.11}$$

2. When sun shines on a square meter of the Earth, the energy in the wave is moving toward the Earth at the speed of light c m/s. It is as if a rectangular solid block of energy $3 \times 10^8 \times 1 \times 1$ m^3 in volume, and filled with energy having density $E_{max}^2/8\pi k$ will pass into that patch of earth every second.

Energy Transmission Examples

1. Question: Given that the intensity of sunlight on the earth is 10^3 W/m^2, what is the energy density, U_{EMavg}, of sunlight?

   Answer: Since U_{EMavg} J/m^3 × 3×10^8 m/s = 10^3 W/m^2, $U_{EMavg} = 3.3 \times 10^{-6}$ J/m^3.

2. Question: What would be the maximum E-field of a light wave supplying the energy intensity of sunlight?

   Answer: Using Equation (8.11), $U_{EMavg} = 3.3 \times 10^{-6}$ J/m^3 = $E_{max}^2/8\pi k$. Therefore $E_{max} = \sqrt{3.3 \times 10^{-6} \times 8\pi \times 9 \times 10^9} = 86$ V/m.

Chapter 9

Light and Optics

Objective: Chapter 8 introduced electromagnetic waves, including light. The emphasis there was on energy— how waves carry energy, reflect off boundaries, and interact with other waves to cause accumulations of energy known as resonances. The purpose of Chapter 9 is to explore how electromagnetic waves carry information we see with our eyes. The wave properties of light cause us to see a reflection at a smooth boundary between different substances. The strength of reflection depends upon the polarization of the light wave. Light also bends when it penetrates a boundary between different materials. A key objective of this chapter is to show why and how lenses bend light waves in a chosen direction to form images we can see. This chapter will permit you to characterize lens images, including their size, orientation, and location.

9.1 Scattering and Reflection from Surfaces

When light comes to a boundary between two materials, the first question is whether the boundary is rough or smooth. See Figure 9.1. If the boundary is rough, there will be a significant amount of scattering, because light waves coming from a specific source toward the boundary wind up going in every random direction.

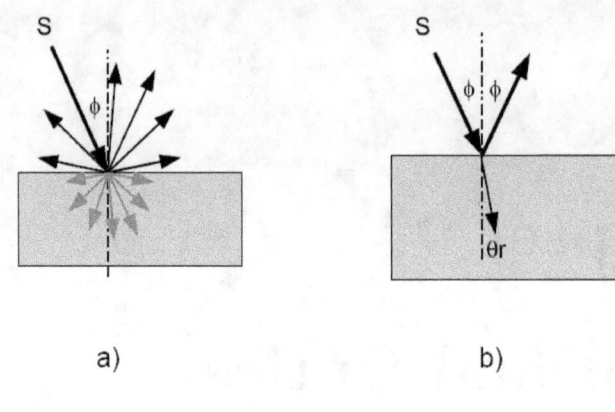

FIGURE 9.1
a) Light from a source S hits a rough boundary between materials, causing scattering. b) At a smooth boundary, there is reflection and refraction of the incident light. The incident angle equals the reflected angle.

If the boundary is smooth, generally two things happen. First, there is *specular* reflection, meaning that we can see an *image* of the source behind the surface. Assume the light is incident on the surface at angle ϕ. The *angle of incidence* is defined to be the angle between the light ray and the *normal* to the surface. Then the reflected ray will make the same angle ϕ with the normal to the surface:

The angle of incidence equals the angle of reflection.

Secondly, there will be some portion of the light that penetrates into the new material. In general, light will have a different speed in the new material. The penetrating wave *refracts*, meaning that its direction is bent relative to the incident ray. The new angle between the bent, penetrating ray and the normal to the surface is called the *angle of refraction*.

When we are looking down into a quiet pool of water during daylight, the situation is exactly as described above. We see an image of ourselves reflected from the smooth water surface. Our image appears to be below the surface looking up at us. Specular reflection preserves all the information in the image, but it is

Light and Optics

located where the light *appears* to be coming from— below the surface. We also see rocks, fish, or weeds that grow beneath the surface. Again, information is preserved, but the image may be shifted because of refraction: that is, the objects appear to be located where the bent, refracted beams are coming from (the angle of refraction), and this is at a little different angle from where the objects are actually located. For this reason, if you angle a straight stick into the water, it appears to have a bend in it at the surface. The stick is not actually bent, but the light coming from the submerged end of the stick has been refracted, and so appears to come from nearer the surface than the stick actually is.

9.2 Metal vs. Dielectric Interfaces

The effects of metal and dielectric materials upon the transmission of light are very different. You recall that metals have a high density of free electronic charge which will respond strongly to the electric and magnetic fields that are carried by light waves. On the contrary, dielectric materials are insulating. They have negative electrons and positive atomic nuclei which can be stretched out of position by the electric force field in a light wave, but the charges all return to their normal positions and give back most of the energy they absorbed from the light wave as it passes by.

Therefore when light hits the boundary between air and a metal, the lightwave E- and B-fields generate electric currents and induce emf's in the metal surface. The currents and emf's in turn generate a new reflected electromagnetic wave and send it back into space as a reflected beam. Any penetrating wave power dies off rapidly in metal, due to current flow power loss, $P = I^2 R$. The light extinguishes completely within a couple of wavelengths of the metal surface. The reflection angle is equal to the incident angle. There is essentially no refracted beam.

When light traveling in air or in free space hits a dielectric medium, little energy is absorbed. Charges are stuck in bonds in the solid, but they vibrate and interact with the E-field in the

Material	Index of Refraction, n
air	1.00
diamond	2.419
cubic zirconia	2.15
polycarbonate	1.58
water	1.33
glass	1.5
leaded glass	1.7

TABLE 9.1: Index of Refraction, n, of some common materials

wave. This abrupt appearance of an electrical response at the boundary of the medium causes a reflected wave which carries back a portion of the incident energy in a reflected beam. Again, the angle of incidence equals the angle of reflection. But much of the energy proceeds to transmit freely through the dielectric medium. Although dielectrics do not remove much power from the wave, they do slow down the speed of light in the dielectric medium. The amount they slow down light is a number called the *refractive index*, or *index of refraction*, usually symbolized by n. The index of refraction is related to the dielectric constant previously discussed in Section 4.6.2.

Table 9.1 gives the refractive index of several well-known materials. Air has nearly the same speed of light as free space, or vacuum, and so air has an index of 1. Diamond is prized for it's brilliance in light, and this is due to its very large refractive index. Diamond bends light through large angles. Cubic zirconia is a cheap substitute for diamond because it has a rather high index, though not as high as diamond. Leaded glass is preferred for crystal glassware and chandeliers because it has a higher index of refraction than regular glass. We are familiar with the light-bending properties of water. Polycarbonate is commonly used to make eyeglass lenses— it has a higher index than glass, so the lenses do not have to be as thick as glass lenses, and they are also a lot lighter in weight.

9.2.1 Refraction and Snell's Law

When a light beam in air strikes a smooth dielectric surface at an oblique angle, there is reflection and refraction. The angle of reflection equals the angle of incidence. What is the angle of refraction? The short answer is that light entering a medium with slower speed of light *always* bends light toward the normal. More exactly, the angle of refraction is given by Snell's Law:

$$n_i \sin \theta_i = n_r \sin \theta_r \qquad \text{Snell's Law} \qquad (9.1)$$

where n_i and n_r are the refractive indices of the incident and refracting media, resp., and θ_i and θ_r are the corresponding angles of incidence and refraction. If the incident medium is air, $n_i = 1$, so Snell's Law reduces to $\sin \theta_i = n_r \sin \theta_r$. Knowing that $n_r > 1$ in a dielectric medium, we have

$$\sin \theta_i > \sin \theta_r \qquad \text{which implies}$$
$$\theta_i > \theta_r \qquad \text{light entering a dielectric from air} \qquad (9.2)$$

This last equation validates our statement that the ray bends toward the normal when it enters a slower lightspeed medium. This was illustrated in Figure 9.1. It is further depicted in Figure 9.2. Figure 9.2 depicts two phenomena similar to light entering a

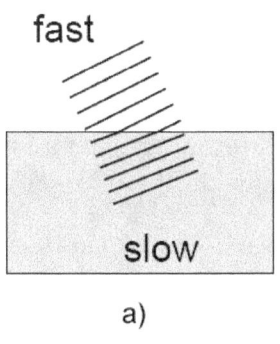
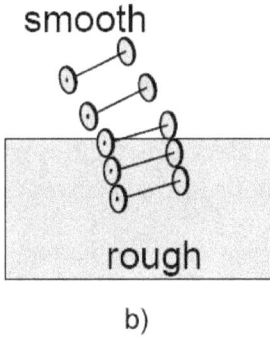

FIGURE 9.2
a) Ocean waves bend toward the normal when crossing a reef where the wavespeed is slower. b) A vehicle turns toward the normal when it encounters a rough patch.

slower medium. Figure 9.2a, similar to a top view of ocean waves crossing a reef, emphasizes the wave nature of light. Parallel wavefronts march through the faster medium. When they reach the slower medium, they must become more closely spaced— i.e., the wavelength becomes shortened in a slower medium because $\lambda = c/f$— Equation (8.4). But because the wavefronts encounter the slower medium at an angle, the left side of the wavefronts is more closely-spaced than the right side. This has the effect of turning the wave into a new direction— namely, bending it toward the normal.

Figure 9.2b shows the top view of a vehicle encountering a rough patch, which is also a slower medium because there is more friction. The left side of the vehicle— i.e., the reader's left— encounters the rough patch first. This slows it down and allows the right side of the vehicle to swing around somewhat so the vehicle points more toward the normal. Both physical situations in Figure 9.2 obey Snell's Law.

Snell's Law Examples

1. Question: A light ray enters a diamond surface at an 79° angle with the surface. What is the refracted angle of the beam inside the diamond?

   Answer: The index of refraction of diamond is 2.42. The angle of incidence with the surface is 79°. Since the index of refraction of air is 1.00, Snell's Law says $1.00 \times \sin 79 = 2.42 \times \sin \theta_r$. Therefore $\sin \theta_r = 1.00 \times 0.98/2.42 = 0.40$. Therefore $\theta_r = \sin^{-1} .40 = 23.5°$. Hence diamond collects light beams from a wide cone of external angles and refracts them into a narrow cone of light inside the diamond.

2. Question: You gaze at a pebble located at the bottom of your swimming pool. Your line of vision is at a 45° angle with the surface. At what incident angle is the light from the pebble striking the surface from within the water?

   Answer: The index of refraction of water is 1.33 and the angle of incidence with the surface is an unknown, θ_i. We

Light and Optics

do know that the index of refraction of air is 1.00, and the angle of refraction up into the air is 45°. Using Snell's Law, $1.00 \times \sin 45 = 1.33 \sin \theta_i$. Therefore $\theta_i = \sin^{-1}(.707/1.33) = 32°$. Hence the pebble sends light up to the surface from a steep angle— 32° from the normal— but you see it at a shallow angle. This means the pebble appears to be in shallower water than it really is.

9.2.2 Total Internal Reflection

An important corollary of Equation (9.2) applies when a light beam goes from a slow medium and enters a faster medium. A beam leaving a dielectric and entering air, for example, bends away from the normal. See the last example in Section 9.2.1. Snell's Law (9.1) still applies, but now we are given an incident angle and refractive index within the slow material, and wish to find the direction in the faster medium. The math is basically the same, but the result is always that a refracted ray leaving a denser material and *entering* air bends *away* from the normal.

What happens when the angle of incidence within the dielectric medium is large? Consider the case of diamond, and suppose two angles of incidence, 24.4° and 24.5° from the normal inside the diamond. The refracted angles should be, respectively, $\sin^{-1}(2.419 \times \sin 24.4)$, and $\sin^{-1}(2.419 \times \sin 24.5)$, resp. The results are 87.9° for the first and $\sin^{-1} 1.003$ (?). Clearly there is no angle whose sine exceeds 1.00, so there is no possible angle of refraction for the second ray! The conclusion is that there is *only* reflection in the second case. Since no energy escapes the diamond when the incident angle is 24.5°, it must be that *all* the energy is reflected and returns into the diamond. This total return of energy, for all angles $\geq 24.5°$ in diamond, is called *total internal reflection*.

For each dielectric medium— including glass, water, plastic, etc.— there is an angle, beyond which the entire light beam undergoes total internal reflection. This has many significant consequences we experience. For example, total internal reflection keeps light moving down a light pipe, such as a fiber-optic, without any loss of energy due to escape from the fiber. This makes fiber optics very efficient for the transmission of light and any information carried on the optical fiber. Total internal reflection also makes leaded-glass crystal and especially diamond extremely bright looking. Any light that enters a diamond tends to be reflected many times within the gem, and therefore exits at a completely random angle. Still another example is the mirage. The mirage is a case where the image from a distant source of light reflects totally when the light skims a hot surface. The hot air layer right near the flat surface of a road or desert has a higher light speed due to higher temperature. Therefore the road surface is like a perfect mirror for light approaching at a glancing angle. The image of distant objects reflects totally.

Total Internal Reflection Example

- Question: What is the maximum angle a ray of light inside diamond can reach the surface and still escape into air?

  Answer: The index of refraction of diamond is 2.42. The maximum angle of refraction for an escaping ray is 90°. According to Snell's Law $1.00 \times \sin 90 = 2.42 \times \sin \theta_{i,max}$. Therefore $\sin \theta_{i,max} = 1/2.42 = 0.413$, or $\theta_{i,max} = \sin^{-1} .413 = 24.4°$. Hence diamond releases light beams from only a narrow cone of internal angles and reflects a majority of the light back inside the diamond.

9.3 Polarization Effects of Reflection

Section 8.3, Figure 8.10, showed the polarization of an electromagnetic wave. The electric field has a specific direction, or polarization, and this affects how the wave behaves when it meets a refracting medium, such as a dielectric. Figure 9.3 shows the two possible polarizations of light incident at an angle to a dielectric surface. In Figure 9.3a, the electric field is indicated coming out of the page. This sets up a vibration of the bound charges in the

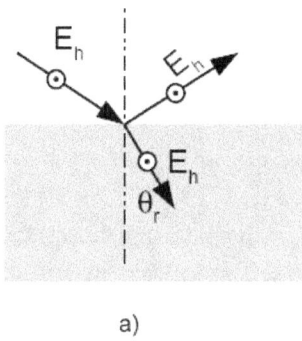
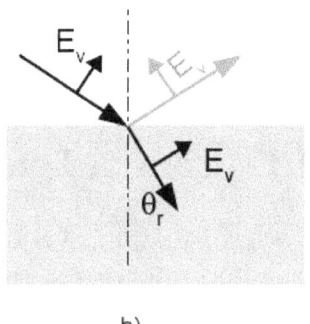

FIGURE 9.3
a) Light polarized parallel to the surface of a dielectric is partially transmitted and partially reflected b) With the polarization in a plane perpendicular to the dielectric surface, the reflection is weakened.

dielectric normal to the plane of the picture. There is no difficulty for part of the energy to reflect, and part to refract into the medium.

In Figure 9.3b, however, the electric field is in the plane of the paper. The refracted beam transmits through the medium without difficulty at angle θ_r. The reflected beam, however, is forced to propagate in the same direction as the incident light's electric field vector. This is not a good match, because the reflected beam must not be a longitudinal wave. That is, it cannot have it's electric field pointing in the direction of propagation. There is a directional conflict between the incident electric field and the reflected electric field. This weakens the reflection. In fact, at a certain incident angle, called *Brewster's angle*— $\approx 50-60°$— the reflected beam is eliminated completely.

The result of all this is that while unpolarized light strikes the dielectric surface at an angle, only the light with E-field parallel to the surface reflects. The reflected light can be 100 % polarized at Brewster's angle. You may have a good pair of polarized sunglasses which will demonstrate this reflection effect from nonmetallic surfaces. When light hits almost any non-metallic surface diagonally— glass, water, painted metal, roadways, etc.— you can rotate your 'Polaroid' lenses through all angles and observe a dramatic drop in reflected light at a specific angle. This makes these glasses very effective for removing a lot of reflected glare.

9.4 Convex Lenses and Images

We have seen in Section 8.3.3 that light waves transmit energy over long distances. In this section we will explore how light projects *information* over distances. Light *scattering* from an interface, discussed in Section 9.1, destroys information, since it mixes light coming from various directions, and re-directs the light in all directions. Refraction and reflection, introduced earlier in the chapter, bend the light in an orderly way. They therefore preserve information.

We have all used a plastic or glass hand lens to focus the sun's rays on a leaf or a bug. Figure 9.4 shows a lens focusing two rays from a distant source, such as the sun. The source is far to the left

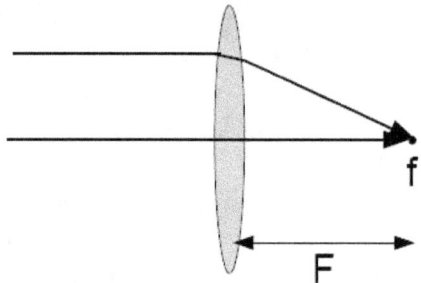

FIGURE 9.4
A convex lens focuses light from a distant source onto a point f.

and does not appear in the figure. Rays emanate from the source in all directions, but it is so distant that rays reaching the lens are virtually parallel. One ray passes right through the center of the lens, and does not bend because it enters the lens normal to the left surface, passes through the glass, and re-emerges normal to the right-hand surface. This is called the *principal* ray. A second ray, parallel to the first, arrives part way up the lens. The lens is double-convex, meaning the front and back surface both curve outward. The upper ray hits the lens at a non-normal angle to the lens surface. Per Section 9.2.1 it therefore refracts away from the normal inside the glass. Thus the second ray bends toward the principal ray. It crosses the lens and reaches the right-hand glass-air surface. The upper ray approaches the second interface at a non-normal angle, and emerges from the glass bending even farther from the normal.

Thus the upper ray bends toward the principal ray on entering the glass, and bends again toward the principal ray on leaving the glass. At a distance F to the right of the lens, the upper ray intersects with the principal ray at point **f**. Point **f** is called the *focus*, and F is the *focal length* of the lens.

Other parallel rays from the source which are farther from the principal ray are refracted more, and those which are closer to the principle ray are refracted less— i.e., through smaller angles. The lens is precisely shaped so that all the parallel rays from the source converge at the single point **f**. This type of lens is called a *converging* lens.

Figure 9.5 shows the focusing effect from the wavefront point of view. Straight, parallel wavefronts approach from the left. Por-

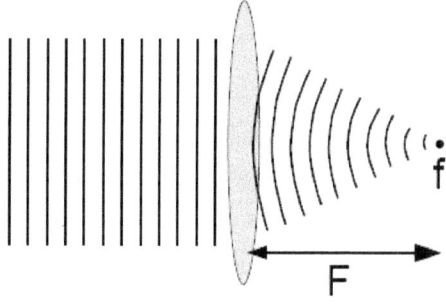

FIGURE 9.5 Convex lens focuses wavefronts from a distant source onto a point.

tions of each wave near the center of the lens are retarded because the center of the lens is thick, and the glass is a slower medium than air. Portions of the wave above and below the principal axis go through less glass and so are less retarded. This gives these outer wavefront sections a little more time to traverse the greater distance they must go to reach **f**. The effect is that all parts of the wavefront reach **f** simultaneously. In accord with our experience, there is a focusing of wave energy at point **f**.

In Figure 9.6a-c, the source of light is moving closer and closer to the lens, until, in c), the source is a distance F in front of the lens. In Figure 9.6a-c, the principle ray, of course, goes through

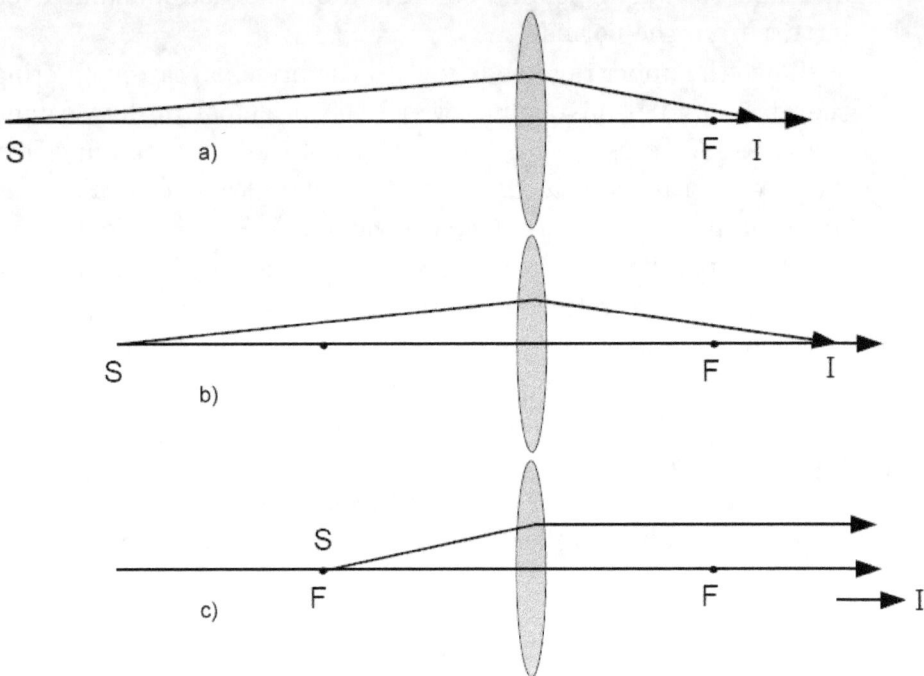

FIGURE 9.6
a) The source moves in from infinity. The image shifts to the right. b) Source moves closer to the lens, image shifts further to the right. c) Source moves to the left focal point, and the image moves right out to infinity.

the center of the lens with no bending. The upper ray now approaches the first glass surface of the lens at a larger incident angle than it did in Figure 9.4. It still bends toward the normal, but not as close to the normal: the angle of refraction must be a little larger because the angle of incidence is now larger. Hence the upper ray bends down at a shallower angle and crosses the principle ray a distance I from the lens, where $I > F$. In fact, the lens is shaped so all the rays from the source will cross the principle ray at the same point. Hence an image of the source forms at distance I.

Figure 9.6c is the case symmetrical to Figure 9.4, where the source was at infinity and the image formed at F. Now the source is at a distance F and the image has moved to infinity. We are assuming the lens is symmetrical, so it must focus a source infinitely far to the right at the same focal distance F to the left of the lens.

9.4.1 Finite Source Size and Distance

Up to this point, the sources and images we considered were all very small points. We want to consider now sources that have reasonable size. We also would like to know the image distance, I, when the source is at a useful range somewhere in between the focus and infinitely far away. Size and distance are available using *ray-tracing*, which is a technique of *geometrical optics*. We trace a couple of rays whose paths we can predict easily; then use geometry to determine the image distance and size.

Figure 9.7 depicts a source whose distance from the lens is S, where S is greater than F. The source is shown as a simplified candle sticking up above the main horizontal axis of the drawing. The figure depicts light emanating in all directions from the candle

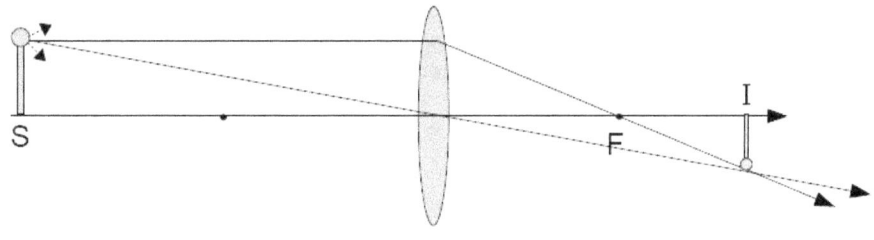

FIGURE 9.7 Geometrical ray-tracing to determine image size and distance, for source distance $S > F$.

flame. Where will these rays be focused by the lens? It suffices to consider merely two rays.

First, we draw a ray from the candle flame through the center of the lens. Although there will be a small jog as the ray passes through the lens, the lens is so thin that the ray passes virtually straight through the center. A second ray is chosen as follows: It should start at the flame, and go horizontally until it reaches the lens. At this point, we know from the case of infinite S that the lens bends a horizontal ray down through the lens focal point at f.

Now extend these two rays to the right until they cross at a distance I from the lens, below the main axis of the drawing. All the rays from the candle flame which reach the lens will converge through this point. This construction determines the location of the image of the flame. Similarly, source points going down the wax of the candle will an image of the wax above the flame image point, nearer the axis. The bottom of the candle will image on the horizontal axis of the figure a distance I from the lens.

There are several important conclusions to draw from the ray-tracing and geometrical construction:

1. The image is *real*. This is meant to imply the image can be viewed by letting it hit a white screen; or can be viewed directly with the eye. Just as rays emanate in all directions from the source candle flame, so also they emanate in many directions from the image of the flame. When we look towards the candle flame, we can see and identify it based upon our understanding of where the light originates. Similarly, we can see the image of the candle flame, and identify its location based upon a similar understanding.

2. The image is *inverted*. Points that are higher up on the candle now appear lower down below the main axis of the figure.

3. The image appears on the opposite side of the lens from the source. The image is reduced in size in the figure. According to the following principle, either *magnification* or *reduction* is possible in going from the source to the image.

4. Any ray tracing can be reversed. That is, we could start with an actual small, inverted candle on the right side of the lens, and trace a couple of light rays from it back to an image left of the lens which would be real, upright, and *enlarged*. Indeed, the arrows in the ray-tracing of Figure 9.7 would be exactly reversed, and the light from the upside-down candle on the right would be projected back to the left side to form an upright candle image that coincides exactly

Light and Optics

with the original source candle shown in the figure. The fact that the rays in a ray-tracing can be reversed is the principle of *reciprocity*.

9.4.2 Image Size Formula

The image size can be derived using similar triangles from your study of geometry. Figure 9.8 repeats the ray-tracing of Figure 9.7. The two grayed triangles in the figure are similar, because they are both right triangles which share included angles. The latter angles are the acute angles formed at the lens, where the diagonal ray crosses the horizontal principal axis of the figure. Because these triangles are similar, their sides are in proportion.

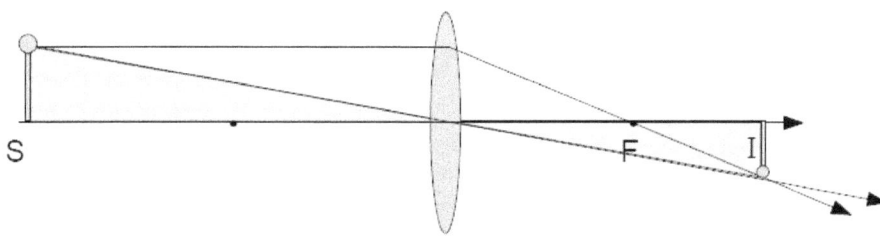

FIGURE 9.8
Figure 9.7 with similar triangles shaded gray.

If we call the height of the source $size_S$, and the height of the image $size_I$,

$$\frac{size_I}{size_S} = \frac{I}{S} \qquad \text{Image size equation} \qquad (9.3)$$

Equation (9.3) states that the image appears magnified according to the ratio of the image distance to the source distance. Both distances are measured to the lens. To determine the image size, we still need to know the image distance. The following section shows how the image distance I depends on the source distance S and the lens focal length F.

Image Size Example

- Question: You are reading a book that is 22 cm tall, and are holding it 40 cm from your eye. Assuming the retinal image is 2.4 cm behind the lens, what is the size of image on the retina?

Answer: From Equation (9.3), $size_I = (I/S) * size_S = (2.4/40) * 22 = 1.32$ cm.

9.4.3 Image Distance Formula

Figure 9.9 repeats the ray-tracing of Figure 9.7. The two grayed triangles in the figure are again similar, because they are both right triangles which share included angles. The latter angles are the acute angles formed at the point where the focused image ray crosses the horizontal axis of the figure at 'F'. Again, the altitudes of the two triangles are equal to $size_I$ and $size_S$. But the two triangle bases are now length $I - F$ and F, respectively. Because these triangles are similar, their sides are in proportion.

FIGURE 9.9 Figure 9.7 with similar triangles shaded gray.

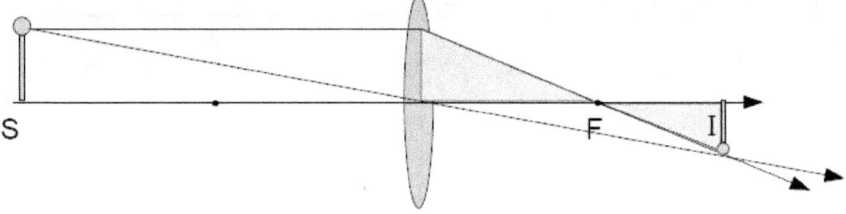

In particular,

$$\frac{I-F}{F} = \frac{size_I}{size_S} = \frac{I}{S} \quad \text{and then dividing by I}$$

$$\frac{1}{F} - \frac{1}{I} = \frac{1}{S} \quad \text{or transposing,}$$

$$\frac{1}{I} + \frac{1}{S} = \frac{1}{F} \quad \text{image distance equation} \qquad (9.4)$$

Example: Finding Lens Focal Length

- Question: A projector lens focuses a slide that is 10.0 cm behind the lens onto a projection screen that is 3.10 meters in front of the lens. What is the focal length of the lens?

Light and Optics   161

Answer: From Equation (9.4), $1/F = 1/S + 1/I = 1/0.100 + 1/3.10 = 10.323$, where all quantities have been expressed in meters. Therefore $F = 1/10.323 = 0.0969$ m.

9.5 Magnifying Lens Optics

Hold a strongly convex lens directly over an object, view the object through the lens, and slowly raise the lens above the object. The object appears larger than actual size. As you raise the lens, the image grows in size until it blurs and fills the entire lens aperture. Raising the lens still further causes the image to reappear inverted. Section 9.4.1 - 9.4.3 studied this inverted, real image and its characteristics. In that section, the image always arose from a source object location outside the focus – ie, $S > F$. The real, inverted image could be larger or smaller than the actual object. The current section treats the magnified image that appears when the object is *between* the focus and the lens itself– ie, $S < F$. Such an image is always larger than the actual source object.

9.5.1 Virtual Image of a Magnifying Lens

Figure 9.10 shows the ray diagram of a source candle placed between the lens and the focus. The figure shows two principle rays emanating from the source flame towards the lens. The lens refracts the rays, but not sufficiently to create a real image on the right side of the lens. Instead, they diverge on the right side of the lens. However, they do form a *virtual* image, in the sense that the diverging rays appear to emanate from the rather large candle image on the far left side of the lens at I.

The image in a magnifying lens is virtual just as the image in a mirror is virtual. No real image is is located 'behind the looking glass'; however, the rays appear to emanate from image points behind the glass.

FIGURE 9.10 A magnifying lens forms a large, virtual image on the same side of the lens as the source object. The dashed lines are backward extensions of the actual light rays.

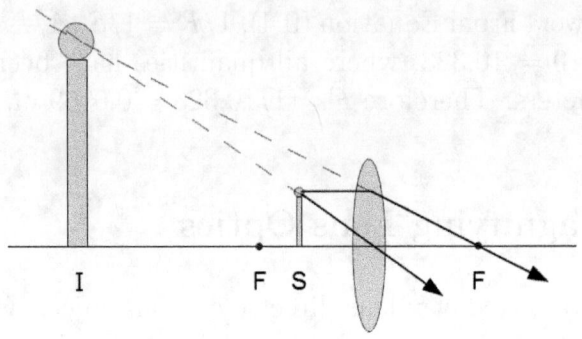

9.5.2 Magnifying Lens Image Size and Location

The geometry steps for finding virtual image size and location again employ similar triangles, as in Sections 9.4.2 and 9.4.3. The ray starting from the candle flame and going through the center of the lens is on the hypotenuse of two right triangles in Figure 9.11: (1) the large cross-hatched triangle and (2) the smaller grayed triangle. These two right triangles share the same included angle between the ray and the horizontal axis, and so they are similar. The ratio of the image size to the object size is therefore still the same as the ratio of I/S, as per Equation (9.3).

FIGURE 9.11 The two overlapping triangles (grayed and cross-hatched) are similar, showing that image size is proportional to image distance.

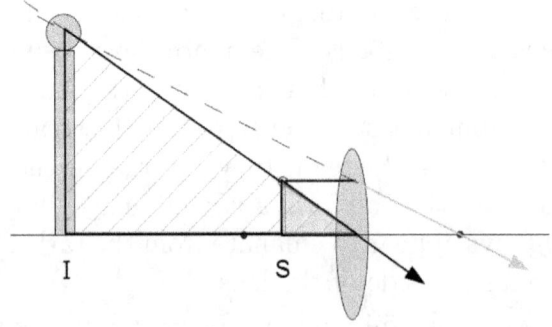

Equation (9.3) for the size of the magnified image therefore applies whether the source is outside or inside the focus of the

Light and Optics

lens— whether the image is real or virtual.

Next, to determine an expression for the virtual image distance, employ triangles formed by the ray that goes from the image to the righthand focus, as shown in Figure 9.12.

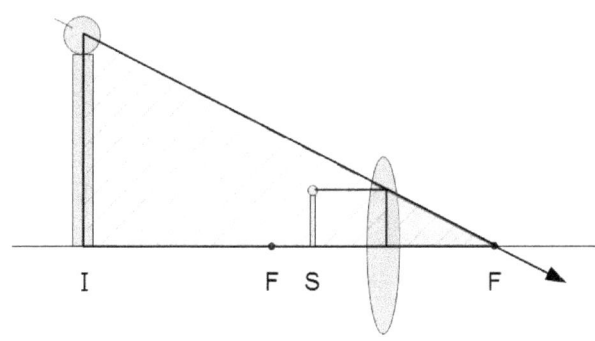

FIGURE 9.12
The two overlapping triangles (grayed and cross-hatched) are similar. This leads to a formula for the image distance of a magnifying glass.

$$\frac{I+F}{F} = \frac{size_I}{size_S} = \frac{I}{S} \qquad \text{dividing by I and rearranging}$$

$$\frac{1}{I} = \frac{1}{S} - \frac{1}{F} \qquad \text{virtual image distance} \qquad (9.5)$$

The above Equation (9.5) is similar to Equation (9.4) except that the roles of F and S are interchanged. When using the magnifying lens, the source distance S is small so $1/S$ is large. It is the sum of the two smaller quantities, $1/F$ and $1/I$.

The following summarizes the characteristics of images produced by a magnifying convex lens:

1. The image is upright.

2. The image is always larger than the object being viewed. If we define magnification, M, to be the ratio of the image size to the source size, we have:

$$M = \frac{size_I}{size_S} \qquad (9.6)$$

3. The image is *virtual*. – it cannot be projected, and the rays do not actually pass through the points of the image, although they appear to do so.

4. Image distance, source distance, and focal length are related by Equation (9.5).

Magnifier Example

- Question: A 10 cm focal length lens is held 8 cm above some fine print which is 2 mm in size. What are the size and location of the image?

  Answer: $F = 10$ cm and $S = 8$ cm. According to Equation (9.5), image distance $I = 1/(1/8 - 1/10) = 40$ cm. The image is on the same side as the source, it is virtual, upright, and its size, according to Equation (9.3), is $(40/8) \cdot 2$ mm $= 10$ mm.

9.5.3 Combining Two Lenses: The Microscope

The optical microscope combines two convex lenses: (1) an objective lens which is carefully positioned over a very small object to be viewed, and (2) a magnifying eyepiece positioned over the real image formed by the objective. The objective produces an intermediate real image of the object magnified M_{obj}, and the eyepiece creates a virtual image of the intermediate image, magnified M_{eye}. This combination of lenses produces an overall microscope magnification:

$$M_{comb} = M_{obj} * M_{eye} \qquad (9.7)$$

A schematic drawing of the microscope showing the ray diagram appears in Figure 9.13. The following is a step-by-step analysis of the microscope design and image formation. Here are the important steps:

- The objective is lowered over the object until the object – say a 'bug' – is just outside the focus of the objective. The objective lens then produces a magnified, real image of the bug.

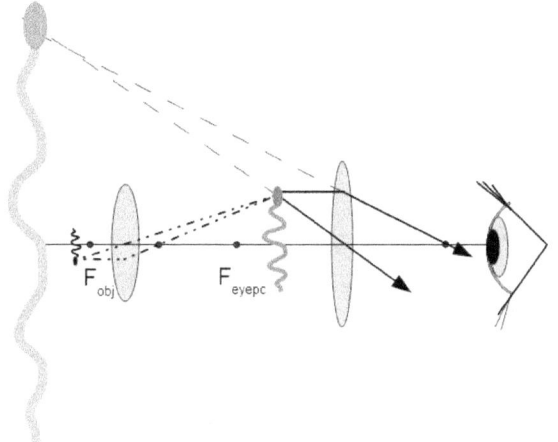

FIGURE 9.13 Microscope Schematic. A 'bug' is placed just outside F_{obj} of a high power convex objective lens. A real, inverted image of the bug is examined just inside the focus of the eyepiece. The magnified virtual, inverted image of the bug appears at the far left.

- The distance between the objective and the bug is critical; it must cause the real image of the bug to locate just under the eyepiece. Therefore the adjustment of the microscope – the 'focusing' operation – is critical.

- The focal length of the objective and the length of the microscope tube – usually 16 cm – together determine the magnification of the real image. Usually the objective is a very powerful lens with a short focal length. The bug has to be near that focus, so the source distance is approximately equal to, but slightly larger than the focal length.

- The magnification is then very large, because 16 cm is much greater than the objective focal length, and so much greater than the source distance. The magnification of the real image is typically between $10\times$ and $100\times$, depending on the power of the objective lens.

- Now the real image of the bug is up near the focus of the eyepiece, which is acting as a magnifying lens. Thus the real image is located just a little inside the focus of the eyepiece. The eyepiece produces a magnified, virtual image of the bug's real image.

- The ideal location for the virtual image produced by the eyepiece is about 25 cm below the eye. This puts the virtual image a comfortable distance, similar to reading distance, below the viewer. Assuming the eyepiece gives 10× magnification, the real image of the bug is therefore 2.5 cm from the lens. The focal length of the eyepiece is just a little bit longer than 2.5 cm.

- The overall magnification of the microscope is typically between 100× and 1000×.

Microscope Example

- Question: A low power microscope objective has focal length 1.0 cm. A biological sample for study lies slightly below the focal point of the objective. The image then forms at the top of the microscope tube, 16 cm above the objective lens. What is the exact location of the sample, and what is the magnification of the image?

  Answer: Applying Equation (9.4), $S = 1/(1/F - 1/I) = 1/(1/1 - 1/16) = 1.067$ cm. The magnification, from Equation (9.3), is $16/1.067 = 15\times$. This magnification needs to multiply the eyepiece magnification, in the next example, to get the the overall magnification.

- Question: The eyepiece lens in the above microscope example has a focal length of 2.5 cm, and creates a virtual image of the sample at a distance 25 cm from the eyepiece. What is the distance from the objective's intermediate image to the eyepiece lens, the magnification of the eyepiece, and the overall magnification of the microscope?

  Answer: Applying Equation (9.4), the distance from eyepiece to the intermediate image is $S = 1/(1/F - 1/I) =$

$1/(1/2.5 - 1/25) = 2.78$ cm. The magnification of the eyepiece, from Equation (9.3), is $25/2.78 = 9.0\times$. According to Equation (9.7), this magnification multiplies the objective magnification to get the combined magnification: $M_{comb} = 9.0 * 15 = 135\times$.

9.6 Diffraction

When light strikes a smooth interface between media, the outgoing light beam is directly related to the incoming light. In the case of reflection, the angle of incidence simply equals the angle of reflection. Information is very well-preserved— as in a mirror image. In the case of refraction, the direction of travel is according to Snell's Law, Equation (9.1). The information is preserved, although lenses can modify the information, as in a magnified image.

Section 9.1 previously mentioned that light striking a rough interface is scattered in a disorderly way in all directions and with total loss of information.

This section deals with an intermediate case between rough and smooth— that is, an interface which has a repeating pattern. The repeating pattern discussed here is a diffraction *grating*. A grating is a series of evenly-spaced dark stripes printed or embossed on a clear glass or plastic slide. This section will explore how a grating bends light rays. This sounds similar to refraction, but the mechanism is very different. In refraction, the change in wave speed when crossing the boundary causes the wavefront to veer into a new direction. In diffraction, the time delay between different openings in the grating causes constructive and destructive interference which sends the light in several different outgoing directions.

Huygen'sPrinciple

Figure 9.14 shows a series of water waves lapping against a narrow gap in a sea wall. The wavefronts are straight lines parallel to the seawall on the ocean side. What sort of a disturbance will pass through the narrow opening into the bay side?

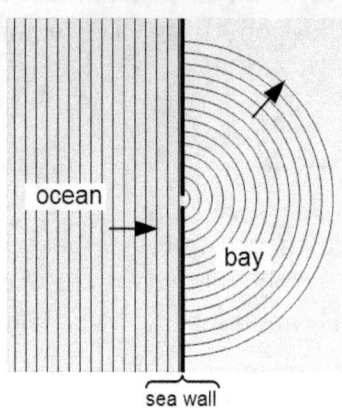

FIGURE 9.14 Waves lap up against the opening in a seawall. Semi-circular waves emanate from the opening into the bay.

The answer is that the waves on the bay side will form concentric semi-circles. This makes sense because the opening in the sea wall is so narrow that waves will propagate outward in all directions as if a narrow paddle pushed the water forward at that point. The wavelength will be the same on the bay side as it was on the ocean side: The frequency of the disturbance is determined by the frequency of the waves on the ocean side— namely, how often waves strike against the opening.

Huygen's Principle states that every wave propagates as if it were composed of a series of many small disturbances along the wave front. The above example of a narrow opening in a sea wall is an example of this principle. In this case, only the disturbance at the opening in the seawall has effect because all the other sources along the wavefront are blocked by the seawall.

As another example, consider a wave with a straight wavefront. If the wave is traveling out on the open ocean, it maintains its straightness. It is as if the whole wavefront were lined up with narrow paddles all pushing forward simultaneously. The semi-circular disturbances emanating from every paddle on the wavefront interfere constructively to maintain a straight wavefront. The sidewards parts of the each semi-circle interfere destructively and cancel completely. Our analysis of diffraction gratings will use

Light and Optics

Huygen's Principle— the idea will be that a semi-circular wave will emanate simultaneously from each opening in the grating.

9.6.1 Two Openings

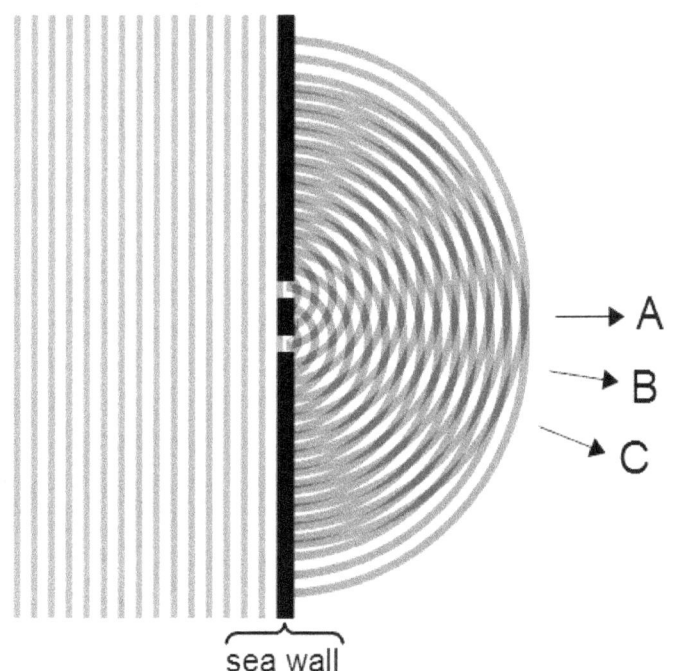

FIGURE 9.15 Waves lap up against two openings in a seawall. Constructive and destructive interference results in waves beaming out in certain directions but not others.

Figure 9.15 shows a series of parallel wavefronts on the ocean side hitting a sea wall that has two openings. On the bay side— the right side—are two sets of semicircles, each centered at one of the openings. Accoring to Huygen's principle, the overall disturbance is the sum of the two disturbances created by the two openings.

Imagine that the trough of a wave is indicated by dark color, and the crest by light color. Then in the straight-ahead direction **A**, the grays overlap to form black, and the whites overlap to form white. This is constructive interference, and the resulting disturbance is twice as big as the troughs and crests created by one slit alone. **A** is the *principle maximum* of the diffraction pattern.

Along direction **B**, the interference is destructive, because the gray troughs from one slit add to the white crests of the other slit, and vice-versa, cancelling at every point. Along **B** there is no net disturbance amplitude. **B** is the first interference minimum. Similarly, **C** shows the direction of the *first order maximum* of the diffraction pattern.

Persons standing at **A** and **C** would experience large waves, while someone at **B** would experience no waves at all.

9.6.2 Multiple Openings

A complete grating has many, many evenly-spaced openings. The diffraction of waves from a series of 24 openings is shown in Figure 9.16. As in the two slit case, the waves from the multiple openings interfere constructively in the directly forward direction **A**. Waves interfere constructively to form maxima at other angles, e.g. towards **C**. Waves interfere destructively to produce relatively smooth minima, such as **B** in between the maxima.

Basically the region directly to the right of the grating, **D**, is a confused region where the different outgoing diffracted beams interact and produce choppy waves. Further to the right the situation clarifies: the beams separate spatially and go off into very different directions of travel. Each direction of travel is perpendicular to a series of straight, parallel wavefronts.

Angles of Diffraction

This section determines the angles of the outgoing beams diffracted from a grating. The principle maximum is straightforward. All the slits emit circular waves in the forward direction. Beyond the region **D** of confusion, the circles are large and unite to form a straight wavefront. This is similar to propagation on an open ocean.

When considering propagation at an angle θ, the circular wave

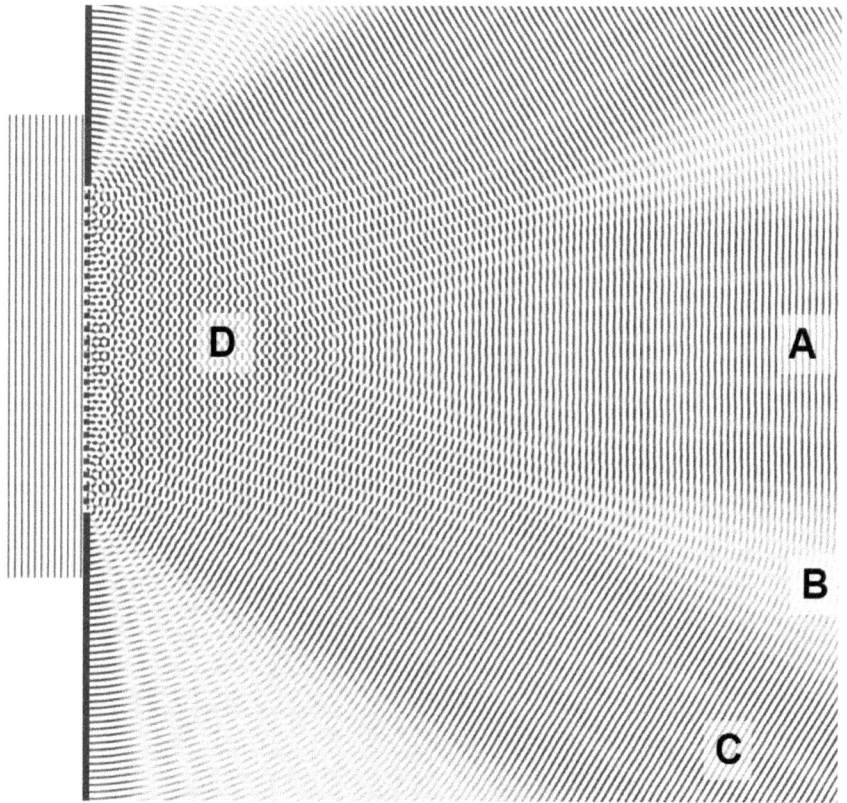

FIGURE 9.16 Waves lap up from the left against a diffraction grating with 24 slits. Constructive and destructive interference appears to the right of the grating.

from each successive slit is delayed a fixed amount because it has a longer distance to travel. This is illustrated in Figure 9.17. The figure shows just three of the slits, and the direction of propagation under consideration is up and to the right in the drawing. Emanating from each slit further down the grating, the wave has to travel an additional distance Δp in order to propagate up and to the right.

FIGURE 9.17 Shows three out of many slits in a diffraction grating. If the wave diffracts up to the right, the paths p_1, p_2, p_3 ..., from each successive slit down the grating is longer a distance Δp.

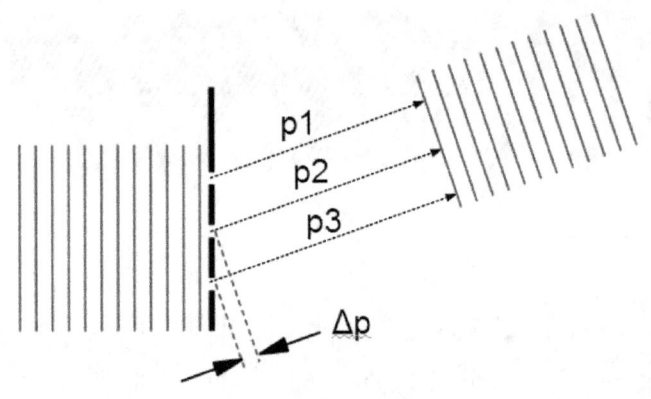

Figure 9.18 shows how to calculate the value(s) of θ for which there are interference maxima and minima. The triangle has an acute angle equal to the diffracted angle through which light is bent. Its hypotenuse is equal to the slit separation. The short leg of the triangle is the path difference Δp between waves from the two slits. At the correct angle for the first diffraction maximum, the waves from the two slits must move up and down together. The path difference must be exactly one wavelength at the angle of the first maximum. We say that they are then moving *in phase*. The equation for the first maximum is

$$\Delta p = \lambda = d \sin \theta \qquad \text{first diffraction maximum}$$

Additional maxima are possible. For example, there will be a second maximum if the path difference can contain exactly 2λ. In general, the complete set of maxima is determined by

$$\Delta p = n\lambda = d \sin \theta \qquad \text{where } n = 0, \pm 1, \pm 2... \qquad (9.8)$$

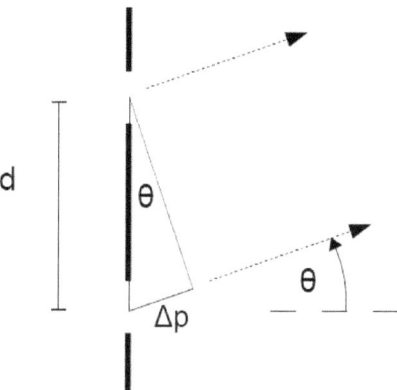

FIGURE 9.18
A right triangle has been drawn between two slits to show the wave path difference from one slit to the next is $d\sin\theta$.

Grating Example

1. Question: Red light wavelength 632 nm is normally incident on a diffraction grating having slit separation $d = 0.0180$ mm. What is the angle between principle maximum and the two first order maxima?

   Answer: To find the first order maxima use $n = \pm 1$ in Equation (9.8). Therefore $\pm\lambda = d\sin\theta_1$. So $\sin\theta_1 = \pm\lambda/d$, $\theta_1 = \pm\sin^{-1}(632 \times 10^{-9} \text{ m}/0.0180 \times 10^{-3} \text{ m}) = \pm 2.01°$.

2. Question: Using the above grating and light source, what will be the angles of second order diffraction?

   Answer: $\pm 2\lambda = d\sin\theta_2$. So $\sin\theta_2 = \pm 2\lambda/d$, $\theta_2 = \pm\sin^{-1}(2 \cdot 632 \times 10^{-9}/0.0180 \times 10^{-3}) = \pm 4.03°$. The diffraction pattern is thus a series of approximately equally-spaced maxima on either side of the principle maximum.

3. Question: What is the angle of first order diffraction when using blue light, wavelength 440 nm?

   Answer: $\theta_1 = \pm\sin^{-1}(440 \times 10^{-9}/0.0180 \times 10^{-3}) = \pm 1.40°$. Hence a diffraction grating bends shorter wavelengths through smaller angles. This is the opposite of the way a prism refracts light. For a glass prism, blue light refracts through the largest angle because the speed of short wavelengths in glass is slower than the speed of longer wavelengths.

9.6.3 Everyday Diffraction

Having defined diffraction and calculated some of its effects, it is good to explore where diffraction occurs in everyday life and science. First of all, through Huygen's Principle, we can see that waves do not travel in straight lines. Clearly water waves diffract around obstacles, but so does light. Light hitting a pinhole passes through and then spreads out— spreads out more the smaller the pinhole. Diffraction gratings have the ability to separate different colors of light. They are very much used in UV (ultraviolet), visible and IR (infra-red) spectrometers, for analyzing the lightwave emissions and absorptions of atoms and molecules. X-rays have a very short wavelength, but the atoms in a crystal are spaced so closely that they behave as a grating for x-rays. X-ray diffraction is a powerful tool for determining the geometrical structure of materials— such as DNA— because every different structure has a unique pattern of x-ray diffraction.

Finally, diffraction gratings can be embossed onto paper and packaging as a decoration. These gratings break up ordinary daylight into its component colors, creating an eye-catching effect of spectral color.

Glossary

Ampere's Law relationship between the amount of current passing through a surface S and the circulation of B-lines of force around the edge of S. 83

angle of incidence the angle between the normal to a surface and an arriving light ray. 146

angle of refraction the angle between the normal to a surface and a bent, departing light ray. 146

antinode location in a standing wave where the amplitude is maximum; a loop. 131

Brewster's angle this is the angle of incidence on a dielectric material which produces no reflection of light having E-field out of the plane of the interface. Only the polarization with E-field *in the plane* can reflect at this angle. 153

bubble chamber a container of liquid hydrogen placed between the poles of a magnet, which records the circular tracks of charged particles from nuclear reactions for the purposes of determining their origin, energy, and identity. 79

capacitance in a capacitor, the ratio between the charge stored and the voltage necessary to keep that amount of charge on the plates. More generally, the ratio between the charge on a conductor and the voltage on that conductor, referenced to zero volts at infinity. 60

capacitor a device designed to store charge, usually made of two metallic plates separated by an insulating dielectric. 55

chord musical mixture of several different pitches created simultaneously. 133

circular wave polarization which is a rotating mixture of horizontal and vertical transverse polarizations. 132

constructive describing wave interference in which deflections add, producing a larger amplitude. 126

converging type of lens that is convex, i.e., thicker in the middle than at the edge, causing a parallel light beam passing through it to converge to a focus. 155

destructive describing wave interference in which deflections subtract, producing a smaller amplitude. 126

dielectric insulating material in which the shifting of positive and negative charges reduces the internal electric field of a capacitor, thereby increasing the capacitance. 63

dielectric constant multiplying factor by which capacitance is increased when a dielectric material fills the gaps between metal plates in a capacitor. 64

domain a small region in a ferromagnetic material in which all the atomic spins are aligned, thus forming a tiny, highly magnetized region. If domains can rotate freely, as in soft iron, the material is easily magnetized in an applied field. If the domains can be aligned and then frozen in position, a permanent magnet results. 88

e the electronic charge, 1.60×10^{-19} Coulomb. 3

elliptical a wave polarization similar to circular polarization, but with one of two perpendicular axes longer than the other. 137

exponential decay time-dependence of a process in which the quantity of interest dies away by a fixed percentage at a time, approaching zero asymptotically. 65

ferromagnetic a class of materials which strongly enhance any applied magnetic field. Iron, cobalt and nickel are ferromagnetic elements. 87

first order maximum the first diffraction maximum occurring on either side of the principal maximum. 170

focal length a lens parameter whose value is the distance between the center of the lens and the image of a distant object focused by the lens. 155

focus state of an image in which the edges, points, and textures are clearly and accurately defined. 155

fundamental the lowest possible resonant frequency in a bounded wave system. 131

Gauss's Law the relationship between the number of E-lines of force flowing out through an imaginary closed surface, and the total amount of charge contained inside the closed surface. 56

GCMS an analytical tool that fractionates unknown materials, esp. organic materials by gas chromatography, followed by identification using mass spectrometry. 79

geometrical optics science of determining image formation by geometrical methods, such as ray-tracing. 157

grating a series of opaque lines and transparent spaces, uniformly arrayed, which bend light by means of interference of the waves passing through the grating spaces. 167

half-life in a radioactive or other decay process, the time it takes for the process strength to reduce by a factor of two. 66

harmonics rational multiples of the fundamental frequency in a resonant wave system. 131

image a replica of a source object, formed in light and appearing in correct proportion, though possibly differing in size, orientation, or left-right handedness. 146

index of refraction when light travels through a material, the ratio of the speed of light in a vacuum to the speed of light in that material. 148

induce to produce an effect on a nearby body: (1) by changing the magnetic flux through a nearby conducting loop, to cause a voltage and/or a current to flow in the loop; (2) to charge a neutral conducting object by bringing into proximity a separate object charged with the opposite polarity. 90

interference when two waves pass through the same point in space at one time, the enhancement or reduction of their combined wave amplitude. 126

Kirchoff's Law circuit principle requiring the sum of the electric currents entering a connection point must equal the sum of currents leaving that point. 40

longitudinal wave mode in which the direction of oscillation of the medium is along the direction of propagation. 132

loop location in a standing wave where the amplitude is maximum; an antinode. 131

magnification the numerical factor by which an optical system mulitplies source size to produce the image size. 158

mass spectrometer an analytical tool that characterizes atomic and molecular fragments by their charge-to-mass ratios, thereby permitting accurate identification of unknown materials. 79

Glossary

node location in a standing wave where the amplitude is zero. 131

normal the direction perpendicular to a surface. 146

note the auditory sensation of a single resonant frequency. 133

octave two pitches differing by a factor of 2 in frequency. 133

period the amount of time between successive peaks when a sinusoidal wave passes a point in space. 124

pitch how high or low a musical note sounds; tone frequency. 133

polarization the orientation of the plane in which a wave medium oscillates; for an E-M wave, the direction in which the electric field oscillates. 132

potential difference the voltage difference between any two points in space or in a circuit; hence the potential energy change of a unit charge moved from the first location to the second. 7

primary one of two coils in a transformer; the primary coil is the input to the transformer. 111

ray-tracing graphical technique of extending a ray path until it encounters an interface between different media, then deflecting the ray path according to Snell's Law. The light is treated as a particle rather than a wave. 157

real referring to an image formed at the intersection of light rays. Such an image can be projected on a screen, and can be re-imaged by a lens systems. 158

reciprocity the principle that the light paths from a source object through a lens system to a real image can be reversed, providing a correct solution for the optical system acting on light going from the image location back to the source location. 159

refract to bend at an angle, usually referring to the bending of a light ray at an interface between dielectric materials. 146

resonance a natural, characteristic motion of a wave system, in which all parts of the system oscillate in unison at one frequency. 130

rms a type of average formed by taking the square root of the mean square of a varying function. The rms value of the sine function is 0.707. 103

secondary one of two coils in a transformer; the secondary coil is the output of the transformer. 111

solenoid an electromagnet formed by coiling a current-carrying wire in a helix. A solenoid can be wound around a ferromagnetic core which enhances the field or serves as a mechanical actuator. 83

specular referring to reflection from a smooth surface in which light rays arrive and leave the surface at the same angle. 146

standing wave an oscillating pattern of interfering waves all having a single frequency. The pattern is fixed in space. It often occurs when a wave reflects between two fixed boundaries. 130

step-down a transformer application in which the secondary voltage is smaller than the primary voltage. 114

step-up a transformer application in which the secondary voltage is greater than the primary voltage. 114

total internal reflection reflection at an interface from a higher index material toward a lower index material, in which there

Glossary 181

is no refracted ray, but the energy is entirely reflected back into the higher index material. 151

transformer an AC magnetic device, consisting of two coils wound on the same ferromagnetic core, used to match the voltage and/or current source capability to the power requirements of a circuit load. 111

transverse describing wave polarization that is perpendicular to the direction of propagation. 132, 134

virtual referring to an image which is *not* located at the intersection of light rays; rather it is located at the backward projection of rays that appear to come from the image, but actually do not intersect. 163

voltage any point in space or in a circuit can be assigned a voltage, which is the electrical potential at that point; it is numerically the potential energy of a unit test charge moved from infinity to that point in space. 7

wavelength space occupied by one complete cycle of a sinusoidal wave, viewed in a snapshot at a single point in time. 124

www.ingramcontent.com/pod-product-compliance
Lightning Source LLC
Chambersburg PA
CBHW080244180526
45167CB00006B/2411